Michael,   May 2021

This is

Henry is a

THE ART OF

CYBER CONFLICT

Based on the book
by Sun Tzu 2000
years ago.

All the Best,

Henry J. Sienkiewicz

Graham

THE ART OF CYBER CONFLICT

First published by Dog Ear Publishing
4011 Vincennes Rd
Indianapolis, IN 46268
www.dogearpublishing.net

ISBN:978-1-4575-5516-9

This book is printed on acid-free paper.

Printed in the United States of America

Cover design by Henry J. Sienkiewicz and Andrew Hiller.

To my nieces, nephew, and cousins

Contents

Introduction

As a senior executive in the federal government and the commercial sectors, a retired United States Army Reserve officer, and an author, I have had the great privilege to lead American infantry and signal soldiers, technologists, and cybersecurity professionals at the highest levels of the American government and in support of efforts across the globe.

While in the federal government, I was personally responsible and liable for the cybersecurity of some of our most crucial military command and control systems. As a corporate executive, I legally delivered technology services to every country in the world. As a military officer, I led both infantry soldiers and cyber professionals—creating doctrine for both. As an author, I have contemplated the way social systems on the micro and macro levels interact.

This book came out of a series of conversations I have had over the years with numerous cyber experts, corporate executives, colleagues, and military professionals. The discussions revolved around the topic of how to think about the problem of cyber, the noun that has become shorthand for a wide range of disparate topics to include the network, systems, applications, and digital devices; conflict in cyber; cyber conflict; cybersecurity; and cyber safety.

I have met some of the most brilliant minds in cybersecurity and some of the most confused minds in cybersecurity. The brilliant minds have thought about technologies and solutions; the confused minds simply want to make their respective organizations safer amid the ever-changing nature of cyber.

What is this book? This book is my response to those conversations. It is an approach to thinking about the complex problem of cyber and cyber conflict. It is intended to give executives and business managers a way to think about the problems surrounding cyber. Concurrently, this book offers the technical team a way to communicate some of the underlying problems in a straightforward fashion.

What it isn't? This is not a technical how-to manual. Technology is constantly changing and evolving. A specific series of detailed steps would quickly be dated.

This book is structured to be highly approachable and provide a pathway for leaders and executives to better reflect and understand the issues surrounding cyber and cyber conflict. It is designed to provide cyber professionals a way to better communicate with their respective stakeholders.

The first seven chapters of this work cover background material on cyber, Sun Tzu, and axioms necessary to recognize the issues. The eighth chapter and its subchapters use *The Art of War* as a framework for discussion and reflection. As the reader will quickly see, Sun Tzu and *The Art of War* can be readily applied to today's cyber challenges.

In this work, each chapter of *The Art of War* begins with "The Stage," which seeks to put a cyber context to the actual text. Next, Samuel B. Griffith's translation of *The Art of War* is provided. Finally, cyber corollaries relating the text to the topic of cyber conflict are presented.

As an editorial note, I have used the public domain version of Griffith's translation. For ease of comprehension, I have opted to use American spelling and remove the traditional commentaries. While the traditional commentaries are of interest, in my opinion, they do not significantly contribute to this discussion. Further, in lieu of other terms, I have used the term *organization(s)* to refer to any individual, corporation, agency, government, and, yes, organization.

Chinese is a language of ideograms, written characters that symbolize an idea or potentially multiple ideas. One of the challenges in translating ideograms is that frequently the symbol can represent multiple ideas. This is evident in the chapter titles. I have taken the liberty of modifying the original translation of *The Art of War*'s chapter titles to better characterize the ideas in relationship to cyber.

My deepest appreciation to my readers who helped me as I struggled through this: Robert Richardson, Andrew Wonpat, Alma Miller, Dorinda Smith, Venkat Sundaram, Adam Firestone, Chris Grady, Andrew Hiller, and Jeffrey Brady.

All of the best,

Henry J. Sienkiewicz

Alexandria, Virginia

CHAPTER 1

Safe

How do I keep my "stuff" safe?

Throughout time and in many fashions, the question of safety has been posed. A parent thinks of safety in one fashion; a police officer in another; a business executive in a third way; a military professional in still another manner.

When you leave your house, you lock your door. When you get in a car, you put your seat belt on.

These are tangible actions that you perform for peace of mind. Underlying each procedure are decades, even centuries, of common practice, regulations, and law. There are building codes to ensure that homes are built safely and securely. There are transportation safety laws and rules that govern both how a vehicle is constructed and how it is operated.

While on a leisurely drive, we do not routinely think about road safety design or car impact zones. Most of us do not lock every door all the time; we keep them locked when we are home mostly for convenience.

We do not, we currently cannot, do this with cyber.

Cyber is the shorthand term for the global physical and virtual electronic infrastructure and its associated interconnected nature, an environment, which has come to define the current age. It is an electromagnetic medium that exists across networks, systems, databases, mobile devices, cloud services, and, most importantly, any electronic network connections. Cyber is that fabric of man-made connections, both virtual or physical.

Safety in the cyber environment *is* different; it is more akin to traveling, to staying at a hotel, than to being at home.

In a hotel, we trust that the individuals dressed like hotel employees or housekeepers are who they say they are. We have no control. The hotel provides them with unfettered access. Our assumption is that they are authorized, especially if they used an approved key.

With cyber, that is not the case. While there are only so many ways to enter our hotel room, with the prevalence of Internet devices, every door is open and anyone can walk in.

In your house or hotel, you would typically notice a theft or malicious activity within a short period of time. You can see if someone has broken in or if something is missing.

In cyber, the thing of value, an organization's data, does not have a physical presence. Intrusions, losses, or espionage can all go on for extended periods without detection—and usually do.

We have assumed that our digital devices are inherently secure much as we assumed that our homes and cars are safe and secure. This assumption is false. The actual risks are numerous, unseen, and intangible, until it is too late.

Mobile phones, desktop computers, laptops, tablets, printers, copiers, smart watches, wearables, home security systems, baby cameras, light bulbs, thermostats, smart homes, drones—all these devices generally referred to as the Internet of Things (IoT)—are all connected to a network; all have operating systems; all communicate; all contain data; all come from different manufacturers, who may have advertently or inadvertently included weaknesses within the devices.

Each one of these devices represents a promise and a liability. While each device provides something useful, each device is also a potential weakness and vulnerability. Such devices promise simplification of our lives through functionality, ease of use, time savings, efficiency, and choice. These devices threaten us with a loss of privacy, piracy, fraud, data manipulation, and disinformation.

These technologies have become pervasive, and former boundaries have been chipped away. For good and for bad, the world has become hyper-connected. An individual can readily connect to almost anyone else anywhere else. A connection can serve to deliver community, news, goods, and services. A connection can also serve as a pathway for loss and theft.

The challenge is concurrently technological, military, political, and law enforcement. Because of its physical and virtual properties, its ubiquity, its unrestricted and deceptive nature, it is difficult to achieve a sense of cyber "safe."

Cyber conflict is not simply a technology issue. With the growing openness and consumerization of technology, the burden of responsibility has shifted. Individuals are responsible for controlling the security of their phones, cars, home appliances, and home monitors just as they are responsible for locking their cars and houses. Information security teams are now responsible for securing their enterprise from devices they are not aware about, do not own, and do not control.

It is not simply a political issue. There are close to 200 nations in the world. Over 120 of these nations actively leverage the Internet for political, military, and economic espionage activities. There are hundreds of thousands of cyber actors. There are domains, boundaries, throughout our world. The boundaries are visible and invisible, military and nonmilitary, state and non-state, corporate and individual. In the cyber conflict, these boundaries have lost their effectiveness.

It is not simply a military issue. The world faces expanding military conflicts going beyond nation state wars into non-state conflicts that are timeless and transnational. Traditional strategic thinking about how to address the threat needs to evolve and, long-standing assumptions need to be set aside in favor of new mental frameworks.

It is not simply a law enforcement issue. Computer and computer-related crime make up a large percentage of all crimes, and the percentage grows every year. However, the legal system has not kept up the rapid changes, and law enforcement agencies have neither the resources nor the expertise to fully safeguard the environment.

Decision makers are generally not adequately educated about the threats, vulnerabilities, or technology. This lack of familiarity, or at least the lack of usefully articulated questions, makes it difficult for them to formulate policy and issue meaningful direction. Except for what they hear in the news, decision makers and executives routinely have only minimal exposure to the threats posed in and by cyber. They can be so bombarded by the noise surrounding cyber that they become overwhelmed, unable to filter fact from fiction, and become desensitized to the threat.

Yet, beneath the noise, certain underlying principles exist. Sun Tzu and *The Art of War* can provide a framework to understand the principles and the complexity of the issues surrounding cyber.

CHAPTER 2
Why Sun Tzu?

So, why Sun Tzu? Can a two thousand–year-old text give insights relevant to modern cyber conflict?

Yes.

In *The Art of War*, Sun Tzu provides a comprehensive, yet flexible approach, an approach that allows organizations and individuals to methodically understand and prepare for any conflict, including cyber conflict. To do so, Sun Tzu put forward six underlying principles that directly apply to the modern cyber conflict:

1. Know yourself
2. Know the enemy
3. Know the environment
4. Use all of your advantages
5. Exploit your enemy's weaknesses
6. Be deceptive and attack along unexpected lines

To fully appreciate Sun Tzu, *The Art of War*, and why it applies to today's conflict-driven world, the historical context of the Chinese Warring States period, 475 BCE to 221 BCE, needs to be briefly examined. The Warring States period was a time of constant conflict and friction among these competing states as they sought to achieve one centralized political empire. As the name implies, warfare was endemic. History documents constant armed struggles and ever-shifting alliances. Conflict was on all fronts.

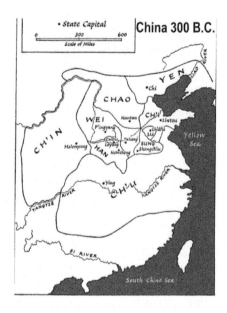

In the previous period of Chinese history, the Spring and Autumn period, 771 BCE to 476 BCE, many of the smaller city-state kingdoms were politically and militarily consolidated into seven major kingdoms. These seven kingdoms dominated the Warring States period.

While tumultuous, the Warring States period was an exceptionally fertile time in Chinese history, rife with dramatic technical, social, political, and military changes. The period is known for its technological innovations and major economic growth. Silk gathering and production, chopsticks, copper coins, acupuncture, and new plowing techniques were just some of the developments of this period. The Chinese game of Go first appeared. Advancements in philosophy and culture include the development of the core Chinese philosophies, specifically Confucianism, Taoism, and Legalism. This period oversaw a broad restructuring of administrative systems. Land and cities were centralized into designated units. (Author's note: While Confucianism and

Taoism are currently thought to be the predominate philosophies native to China, Legalism, the "belief that human beings are more inclined to do wrong than right because they are motivated entirely by self-interest," became the official philosophy of the era immediately succeeding the Warring States period.)

Significant changes revolving around the way warfare was conducted also occurred. During the Spring and Autumn period, most battles were carried out by small groups of chariot riding nobles. The use of a sword or bow and arrow required many years of training to produce lethal effects. The costs incurred and training required meant there were relatively few qualified men to lead.

During the Warring States period, military technology advanced. This progress included the individual casting of weapons that substantially improved the armament of the foot soldiers, the use of mass groupings of foot soldiers rather than chariots, the development of the crossbow, the enhancements that changed the *jian* from a defense pike to a more offensive sword, the perfection of iron casting and forging, the initial construction of the Great Wall, improvements to the war chariots, and the use of cavalry.

Amid all these changes, Sun Tzu wrote about warfare to train others to understand the changes and lead military forces in this new environment. While he remains one of the most famous military theorists in Asia and more recently in the West, little is known about the man; even his name is ambiguous—*Tzu* being an honorific term meaning *master*. What we do know is that he was born in the state of Qi and served the king of Wu, Ho-lu, during the Spring and Autumn period.

Since Sun Tzu lived in the Spring and Autumn period, there is a great deal of scholarly debate as to who wrote *The Art of War* and when it was written. It was a common practice within Chinese historical and philosophical writings to credit authorship to historical figures to give the text more credibility. Nevertheless, conforming to standard practice, Sun Tzu will be cited as the author throughout.

Most historians date the actual writing of *The Art of War* to the Warring States period. *The Art of War* specifically references military techniques and equipment that were not available during the Spring and Autumn period. Further, passages such as Chapter 8, subchapter 5, "Force," are grounded in a Taoist philosophy that only became common during the Warring States period.

Current changes, conflicts, and technological advances have created conditions today similar in nature to the conditions during the Warring States period. It is the context of change, conflict, understanding the environment, and technological advancement that allows *The Art of War* and Sun Tzu's principles to be used as a framework for understanding cyber conflict. This begins with an understanding of cyber and the cyber environment.

CHAPTER 3

Cyber

Historical Perspective

W hat is this thing called the cyber? And why does it matter?

It matters because it has become the global nervous system. People connect; business is transacted; nations are secured, all on this common, electronic platform.

This interconnected electronic infrastructure began as a simple way for a small, decentralized, trusted community to readily and easily share information. One device would connect to another; they would exchange an electronic handshake and conduct a transaction. No questions were asked. The larger the number of connected devices, the better and more efficient the information sharing. This interconnected community of academic and military users generally knew and trusted each other. In those early days, there was nothing but rave reviews of how this new environment would beneficially change the world. This environment was thought to be a cure-all.

Steven Levy's

Hacker Ethic

1. Access to computers—and anything that might teach you something about the way the world works—should be unlimited and total. Always yield to the hands-on imperative!

2. All information should be free.

3. Mistrust authority—promote decentralization.

4. Hackers should be judged by their hacking, not bogus criteria, such as degrees, age, race, or position.

5. You can create art and beauty on a computer.

6. Computers can change your life for the better.

This garden was the combination of a utopian vision, libertarian economics, and a Westernized open society. Steven Levy's *Hackers: Heroes of the Computer Revolution* (1984) provided an early manifesto for this cyber garden under the label of a "Hacker Ethic." As seen in many publications and on many websites, this manifesto has six principles, principles that have achieved an almost spiritual property.

By the mid-1990s as the Internet spread with this presumption of free sharing and access, the new *cyberspace* acquired almost mystical properties. Technology's evolution changed the previously paper-based world into a digital world of interconnectivity and dependency, a cyber world.

An early Internet philosopher, Michael Hauben, coined the term *Netizen*. The term initially had two meanings. As Hauben described it, "The first is a broad usage to refer to anyone who uses the Net, for whatever purpose ... The second usage is closer to my understanding ... people who care about Usenet and the bigger Net and work toward building the cooperative and collective nature which benefits the larger world." The term grew to encompass the idea of a "citizenship" of sorts, one in which national sovereignty was replaced by a network-based, virtual community.

Since those early days, the world has become more complex, connected, and continuous. The individuals and devices are now highly diverse. The individuals and devices are all electronically tethered together. The individuals and devices are always on.

Today, data travels on undersea cables, through satellites, and across shared spectrum. Families and friends use these pathways to connect. Companies use this infrastructure to transact business, exchanging trillions of dollars every day.

When the networks go down, connections ceases. Friends can no longer reach each other. Business and financial services do not slowly wind down; they abruptly stop. For industries in which millisecond latency costs millions of dollars, the economic impact of network slowdowns or outages is highly disruptive and next to impossible to quantify.

While the technology that society relies on started out as government funded, it is now based on research carried out in universities and by private companies; it is mostly developed and owned by the private sector. Technologies once thought to be purely for government and military purposes are now adopted for personal, commercial, and entertainment uses and vice versa.

The harsh, competitive reality of the marketplace and competing politics has caught up and overrun the Robin Hood, altruistic mentality that originally prevailed. Far from being Levy's utopian vision, this environment is actually dystopian and hostile.

Levy's and Hauben's assumptions have proven naive and, on some level, inadvertently deceptive. Levy initially did not consider that cyber is completely man-made, -owned, -maintained, -updated, and -monitored. Each piece is created, owned, and licensed by someone or something, including the pieces released under the various open-source licenses. The design principle that cyber actions and transactions are fundamentally open and trustworthy have been proven false.

Further, the political climate around this always-on connectivity has changed. The electronic world is becoming bordered and moving away from Westernized civil society control.

Cyber Environment

With multiple electronic, connected devices throughout homes and cars, and within every pocket, the small, trusted electronic community is clearly no more. While there is no longer any land being built or seas created, the cyber environment continues to grow and expand.

Cyber safety is not about the adversary. The same techniques that protect against credit card scammers in Chennai will protect against political information hackers in Russia.

Cyber safety is primarily an environmental thing. While actors matter, cyber conflict revolves around safeguarding data in a hostile environment. As with ocean navigation or space exploration, the successful navigation of the cyber environment requires preparation and precautionary measures in order to protect the cargo.

To be explicit, a cyber target, the thing of value, is the data. The cyber environment is comprised of the "things" surrounding, storing, safeguarding, and transporting the data.

Malicious actors want data. They get to that data by maneuvering through the terrain, the environment.

To modify the definition put forward by the United States military, the cyber environment is composed of "those physical and logical elements of the domain that enable mission essential (warfighting) functions." While similar to the purely physical terrain described by Sun Tzu, cyber terrain has some significant differences due to its simultaneous physical and virtual nature. It is the combination of those natures that makes identification of and operations around cyber terrain more difficult than the other traditional military domains—land, sea, air, and space.

The cyber environment is tied to physical devices that have logical locations. The physical component is composed of the infrastructure's hardware. These are the devices that organizations use to connect and move data around, to include the networking equipment, intrusion detection systems, data servers, and cable plant. It also includes the hardware not traditionally under the control of any single organization; the electrical grid, the Internet providers' switching equipment, the telephone company's central office, and the water supply systems are all examples of the greater cyber environment.

This is concurrently a physical and a virtual environment, where physical actions impact virtual actions and vice versa. The virtual nature of the cyber environment allows for the dynamic creation, modification, and destruction of the terrain. Defense has become a matter of making an asset value assessment and then allocating defensive resources more heavily to safeguarding the highest value organizational assets. In some constructs, such as the immutable infrastructure concept, defense is predicated on regular cycles of destruction and regeneration of assets. While the cost to retrofit defense may be high, the cost of built-in, integrated defense may not be.

To find the most effective approach, there has been an attempt to force cyber as a domain to directly fit into the military or law enforcement models. These models have applicability. The problem is cyber conflict is inherently different. By its very nature, it is "civilianized," directly residing within most people's daily lives.

Cyber Sovereignty

The cyber infrastructure is simply a collection of interconnected but independent networks. This architecture allows governments and major organizations to create physical network choke points and gateways, resulting in an evolving notion of cyber sovereignty, a "cyber Westphalia" of bordered geopolitical and geo-economic jurisdictions.

The term *sovereignty* refers to the right of a governing body to rule itself without external interference. As such, a government possesses supreme power over some polity. In the realms of international political systems, the term *sovereignty*, specifically *state sovereignty*, denotes recognized authority over a given population in a defined state by a supreme authority.

The notion of sovereignty dates to the classical rule of the Roman Empire and has evolved over the years. The Peace of Westphalia was a series of peace treaties that ended the Thirty Years' War (1618–1648). These treaties established the legal basis for national self-determination, in which each nation-state has dominion over its territories.

As a concept, sovereignty is the cornerstone of international law. It provides the basis for who has final authority on the internal matters of a given state. More importantly, in this context, sovereignty is the legal basis for recognition by other nations. Sovereignty defines a state's internationally recognized borders; it affords nations specific rights in international legal bodies should disputes arise.

Cyber sovereignty is the notion that governments can exercise control over their digital environment, *however* they define it. While it is not easy to pinpoint data ownership or origination, it is relatively straightforward to see data as it resides within a region and when data is in transit between locations. For the servers, hosts, and data at rest components, cyber sovereignty can be very clear. Rights to the data reside with the nation where the servers physically reside. For data traveling from one country to another, the control can be defined by laws and jurisdictions negotiated among the respective parties. Cyber sovereignty is keeping an eye on and controlling these Internet-based activities, especially financial, political, and security/conflict activities.

For financial activities, governments exercise supervisory control. From regulating monetary transactions to prohibitions on specific activities to preventing scams to imposing taxes, governments have a legitimate interest in safeguarding the financial activities of its citizenry.

For political activities, fundamentally, this environment is not devoid of politics. Rather, it provides a platform for politics. The motivations driving governments all over the world to exercise power over the Internet are clear. Important political issues and messages are often controlled. This can manifest itself in many forms, including efforts to censor political opposition, exposure of corruption, criticism of the military, disparagement of the ruling families, and "blasphemy."

Incidents of governments legally, although not necessarily morally, exercising control in the cyber domain continue to increase. Border agents across the globe demand access to social media and e-mail accounts. The Chinese government has made clear its intentions to have complete control over the cyber activities within their borders. In Brazil, a Facebook employee was sentenced to prison when he refused to hand over data to the government. In Ethiopia, the government shut down the Internet to prevent negative publicity due to the then prevailing drought conditions. The Vietnamese government blocked Facebook to stop protests against environmental damage. Countries like the United Kingdom and Russia have established laws that give their governments special powers to maintain a watch on Internet activities within their borders. Customs agents in both the United States and Canada routinely examine travelers' electronic devices.

Governments have developed many ways of handling cyber sovereignty and the related issues. The method employed depends on a variety of factors. Some issues can be handled with a request from the government to the author of the content to remove it. Another very effective method of enforcing control over the Internet is to have intermediaries to exercise control and increase their surveillance over third-party content that is published on their websites. This latter method is quite effective, as this distributes responsibility. The intermediaries deal with the third-party content as they deem fit. In cases where the softer means do not work or the content is considered too great a threat, the government can resort to actions like forced removal of content or arrests.

Cyber Conflict

There is valid confusion regarding the *"who," "what,"* and *"where"* of cyber. Akin to the debates within philosophical and theological circles, there has been enormous discussion to what constitutes conflict in this new arena. While it is emerging, there is no current, widely accepted lexicon. Terms such as "cyberconflict," "conflict in cyber," or "cyber conflict" abound, each with its own nuances and strong adherents.

Not to overly simplify the discussion, cyberconflict has been defined "as the use of computational means, via microprocessors and other associated technologies, in cyberspace for malevolent and/or destructive purposes in order to affect, change or modify diplomatic and military interactions between entities." Actions occur solely within the virtual domain.

This limitation does not recognize the underlying reality. The actions within the virtual domain impact the physical domain, and vice versa. "Conflict in cyber" characterizes activities occurring in either or both the physical and virtual domains. For ease of conversation, "cyber conflict" encompasses this broader meaning.

It is natural in the exercise of sovereignty in both the physical and virtual worlds that disagreements and frictions arise. *Conflict* refers to some kind of friction, disagreement, or discord that arises between or within groups. Political, military, or economic conflict has been defined as the contest or struggle between people having opposing ideas, beliefs, values, needs, or goals. By this definition, conflict is not always characterized by violence. However, conflict's actions might lead to destructive results.

The modern post-bipolar era is characterized by two seemingly opposite trends. On one end of the spectrum is that the risk of a world war or any open conflict between the world's major or influential countries is very low. On the other hand, there has been a dramatic increase in the number of regional, interstate, and intrastate skirmishes.

In this modern world, the nature of conflict has changed. Recent literature indicates that there are approximately fifty countries in the world that want their neighbors' territory. New actors have emerged, actors who do not adhere to the traditional "laws of land warfare" or the international conventions governing conflict. These new actors include multinational organizations, transnational armed forces, criminals, and terrorists. The international community and law have struggled to resolve all the issues that are caused by these players.

In popular culture, cyber conflict invokes a glorified image of a lone actor standing up to some maleficent corporation or intrusive government—like the *Matrix* trilogy or the television show *Mr. Robot* or the image of a nation averting a massive attack through the strategic emplacement of a computer virus that crippled another nation. Or the image of a disheveled, unshaven person sitting in a dark room in Eastern Europe, or the mask of Guy Fawkes representing the online hacktivist network Anonymous.

The reality is grayer and grimmer.

Traditional military engagements have been replaced with smaller, incremental events designed to collectively deliver the desired outcome. Traditional law enforcement practices have proven inadequate.

Cyber conflict, like conflict in general, has been described as a "war of a thousand cuts." The cumulative impact of these cyber cuts on thousands of small business, websites, e-mail accounts, social media, mobile devices, electronic commerce, and individual websites is uncountable but enormous.

This is a world of rogue hackers stealing information with the intent of selling it. Compromised devices connecting to the corporate networks and siphoning off research; unknown contractors walking through offices, tapping into databases; cybercriminals placing hacking devices under cash registers.

Cyber conflicts can involve the use of simple actions like an e-mail request to send money to sophisticated encounters between or among hidden and distributed networks or personnel. Virtual armies can line up facing each other on a digital battlefield.

As will be seen, traditional military theory has four challenges: identification, location, execution, and results. These challenges once had defined answers. The enemy was clearly identified; the terrain did not move; the results could be readily measured; the pace of execution was measured. This is not the case in cyber.

Cyber actors are difficult to identify. Cyber terrain is highly dynamic. Results can be difficult to attribute and to measure. The pace of execution has an atemporal nature to it, concurrently rapid and slow.

Often, cyber actors are not military in the traditional sense. International law has established four common attributes of a legally recognized military that set the internationally recognized military forces apart from irregular forces. Cyber actors do not traditionally adhere to these four characteristics. The attributes defining a recognized military are:

- An organized military body

- A hierarchical structure answerable to the highest levels in the entity or the state

- A legal status to bear arms and to have separate disciplinary code

- Centralized funding for the purchase of warlike material

The actors are highly active adversaries; they are creative rivals and not passive targets. Attacks are disguised by deception. Incursions are delivered with blinding speed to unexpected places. Cyber actors are always adjusting and reacting, with no intention of adhering to someone else's plans.

The world has changed. This is now an era of "cyber conflict," a physical *and* virtual conflict among states and non-state organizations. Success in this conflict requires a cognitive framework that uses attributes from traditional military and law enforcement theories, and from Sun Tzu to understand the cyber actors and environment.

CHAPTER 4

Complementary Theories

Technologists, businesspeople, and the military commonly share one attribute—they like to get things done. They have a strong preference to action, preferably bold action. They ask *what* and *how*.

With an operational focus, *just get it done* is their motto. They generally do not like to ask *why*. Things are done regardless of whether a strategy exists as to why one is doing something in the first place.

The complexity of cyber precludes a simple categorization with a single, simple, grand theory to explain the *why*. Two existent categories of theories provide a great deal of insight on the way to approach cyber conflict: military theory and law enforcement theory. However, neither one individually nor collectively provides the whole solution.

Military theory pertains to safeguarding sovereignty from external threats, while law enforcement theory applies due to its focus on internal safety. Leaders, managers, and practitioners need to draw from these theories in order to safeguard the data and the environment.

Military Theory

For the domains of land, sea, and air, theorists have written extensively in order to provide insights. A measure of the durability of the theoretical insights of military strategists and theorists like Carl von Clausewitz, Antoine-Henri Jomini, Alfred Thayer Mahan, Giulio Douhet, Hugh Trenchard, William

"Billy" Mitchell, and John Boyd is that their insights transcended the traditional models of warfighting to focus on an underlying theory of war.

As a bridge between the gap of other military theorists and cyber, Boyd's insights were about continuous information gathering, sense-making, course of action development, and execution. In essence, he was all about the *why*, which leads to the *what* and *how*.

Current military practice has five operational domains: land, sea, air, space, and cyber. Traditional military theory for land, sea, and air addressed four themes: identification, location, results, and execution. The two newer domains, space and cyber, have not been addressed with the same rigor.

For most of history, the answers to these challenges were straightforward. The enemy was clearly identified; the terrain did not move; the results could be readily measured; the pace of execution, while a key variable, was controllable.

First, the enemy could be clearly identified and known. A state of war was declared. A general overlooking the battlefield could clearly distinguish the enemy's line. Deception and anonymity were tactics, not inherent aspects of the battlefield, which is not the case in cyber conflict.

The battle's location was established. The general could march his forces to a point where the next day the battle would be joined. The roads did not shift, the hills did not move. The physical terrain had permanence.

In these battles, the lines were drawn and visible. It was evident when and where the one side was wining and the other side was losing. Results could readily be measured, usually visually.

Traditionally, the pace of execution was measured in the time it took a physical object to reach the intended target. Armies took time to march to their objectives; ships and submarines had to cross oceans; aircraft had to cross the skies.

The physics of motion dedicated the pace of execution. Up to the twentieth century, speed, distance, and, for Boyd, the ability to observe and make sense of the environmental conditions were key limiting factors in conflict. Even today, the most advanced warfighting units anywhere in the world can take up to eighteen hours to arrive. In cyber, one click and in 0.5 seconds a force can be projected across the globe.

To understand cyber conflict is to understand the challenges the environmental changes pose to the traditional approaches to these four themes. In cyber conflicts, the enemy is *not* clearly identified; the terrain *does* move; the results cannot be readily measured; the pace of execution is measured in *milliseconds.* The impacts can be disguised as another technology failure rather than the result of a cyber operation

The view that cyber conflict can be executed as traditional clash of wills through decisive offensive and defensive actions is mistaken. A response to cyber conflict is categorized by nontraditional thought models, acceptance of widespread deception, acknowledgement of missing boundaries, realization of universal agency, and a willingness to accept thousands of "cuts." An understanding of this new environment and responsibilities can be found in *The Art of War* and Sun Tzu's philosophy.

Law Enforcement Theory

Crime is often highly opportunistic, when actors take advantage of a readily available circumstance. Actors choose targets that require little risk and effort but offer high reward. Criminal go where they can get the most for the least. Cybercrime is no exception to this rule.

With its focus on internal safeguards, law enforcement theory complements warfighting theory's focus on external security. Two law enforcement theories provide insights into safeguarding the cyber domain—the broken windows theory and community policing.

Broken Windows Theory

In criminology, Drs. George Kelling and James Wilson proposed the broken windows theory in 1982. According to Kelling and Wilson, "Broken windows is a criminological theory of the norm-setting and signaling effect of urban disorder and the impact of vandalism on additional crime and anti-social behavior." According to the theory, a building with unrepaired, broken windows will produce an inclination for other crimes to occur. The community norm is disorder, which, in turn, breeds crime.

The model takes a preemptive stand toward offenses and crimes to prevent crime from occurring. (Kelling, 1996) argued that disorder is not directly linked to serious crime, but instead, disorder leads to an increase of fear and sometimes withdrawal from residents, which ultimately allows more serious criminal activities to move in because of the reduced levels of informal social control.

Similarly, an unmaintained, unpatched cyber environment becomes a potential target, and poor internal practices become the organization's cultural norm. Both can lead to the conditions that allow for a successful cyber attack.

Community Policing

In the physical world, community policing restructures police organization and daily routines to focus on "community" engagement. Community policing is a philosophy of policing based upon the idea that law enforcement must have a more in-depth understanding of the public in general and the specific community it serves. The occupying force model is rejected in favor of a model within which officers collaborate with the general public. Law enforcement cooperates with the general public as the police seek to understand what are the environment and the community norms.

Continental European countries have embraced community policing, often re-branding it as "de proximité / proximidad," or neighborhood policing (Dupont 2007, Emsley 2007). The primary focus in continental countries is on location and neighborhoods, with policing methods supporting a centralized and state-centered tradition.

Non-Western regimes have adopted at least the language of community policing. In China, the emphasis is on the collective responsibility of citizens to the State and the maintenance of order. "Safety and security are maintained through native social and political structures that demand conformity to the collective and to the State" (Wong 2001).

From its onset, cyber was envisioned as a community. As the environment has grown, the tensions and frictions analogous to those of physical communities have carried over. Community cyber policing, through alliances, partnerships, standards bodies, and third-party testing, can assist organizations in ensuring the safety of their environment and the establishment of community norms and standards.

* * * *

From the four military principles of identification, location, results, and execution to the need for the community to look after itself with law enforcement's assistance, military and law enforcement theories provide a historical reference model for cyber. The difference is that while the military and law enforcement are organized, have a hierarchical structure, and have a legal status to bear arms, cyber actors frequently do not have organization, structure, and status.

Yet the cyber environment is everywhere, and cyber actors can reach into every living room, every boardroom, and every device. With this pervasiveness comes the need for great responsibility.

CHAPTER 5
Agency

First and foremost, where the response to cyber conflict significantly differs from tradition warfighting and law enforcement is that everyone is a potential actor; everyone has responsibilities; everyone has agency. Cyber safety is not the sole responsibility of a technical team. Individuals and organizational leaders are primarily responsible for their cyber safety.

Agency is commonly defined as the capacity of an individual or entity to act in a given environment. The actions can be either purposeful, goal-directed activity or unconscious, involuntary behavior. Fundamentally, cyber actors have agency—they know what they are doing.

Agency is locking the front door and turning on the home alarm system. Agency is wearing a seat belt and routinely maintaining the vehicle.

The ubiquitous nature of Internet devices, from computers to phones to cars to home-monitoring systems, means that everyone everywhere can act and interact on the Internet. Measures to ensure cyber safety are required within the office, the home, or the car. Everyone has become an actor, an actor with agency and responsibility.

Responsibility

From the amateur to the sophisticated and professional, coordinated cyber attacks can wreak havoc. These attacks are not the result of conventional, nuclear, or biological weapons or an invading army. These attacks use commonly available, unspecialized equipment, often consisting of readily available laptops, modems, telephones, and software to prevent access to essential services like electricity and water.

While the definitions of cybersecurity varies from country to country, cybersecurity can be summarized as the collection of equipment, guidelines, policies, training, personal awareness, assurance, best practices, and risk-management actions and approaches used to protect the cyber infrastructure, organization, and individual users' / communities' assets. In a world where bank services, elections, shopping, and even parking payments are done through the Internet, everyone has cyber responsibilities—governments, organizations, and individuals.

Governments

For governments, there are four major responsibilities:

- Develop and take some countermeasures to thwart cybersecurity threats by creating common awareness and understanding of cybersecurity in all the stakeholders, establishing policy at a high level and coordinating national action. The cybersecurity policy should highlight the importance of cybersecurity to the nation and identify the potential cyber risks and threats.

- Stop its citizens' involvement in cybercrime via proper legislation. Criminal law is established to ensure the prevention, investigation, and prosecution of all forms of cybercrime. The laws should be aimed to provide security in electronic communication, prevent fraudulent use of computers, and promote the protection of personal data and privacy.

- Test the readiness and response of the incident management teams by conducting cybersecurity exercises.

- Create awareness in end users through generally available training programs.

Organizations

Cybersecurity is not only the responsibility of government but also of industry, firms, and business organizations. Cybersecurity is a leadership issue; it is not a responsibility that can be delegated to a technical team. Executives and boards are the ones ultimately responsible for their organization's cybersecurity. Firms have a positive legal obligation to take two broad types of precautionary measures:

- Implementing security measures to protect themselves and their stakeholders.

- Ensuring that the products delivered have built-in cybersecurity provisions.

Individuals

As the technology advances and computing becomes ubigitous, the role of the individual has shifted. Individuals need to be recognized as the most influential actors in securing cyberspace. Every time an individual goes online, he or she leaves a digital trail behind, a digital trail that encompasses everything an individual does and a digital trail that cannot ever be deleted. Either intentionally or unintentionally, willingly or unwillingly, the vision espoused by Hauben has been achieved; all individuals have become digital citizens, actors within cyber. With citizenship comes rights and responsibilities.

Cyber Citizens

Cyber citizens are anyone or anything—individual, self-motivated group, corporation, government—accessing the infrastructure or data, with or without explicit permission.

Information security professionals are familiar with malicious cyber citizens (i.e., the actors who seek to gain unauthorized access to enterprises' computer networks), but that definition is often too broad to have much utility.

Often, malicious cyber citizens are outside the direct control of a recognized nation-state. Not responsible to anyone, they are not limited, restricted, by the traditional laws of warfare. They engage in activities that go beyond traditional norms for nations. For these actors, the historical prohibitions on nonmilitary targets are gone, with the opponent's weaknesses creating a prioritized targeting selection criteria.

From the lone wolf to a hostile nation-state, the adversaries, cyber citizens, need to be understood and appreciated. In developing an able defense, one must understand that there are a wide variety of motives and capabilities, ranging from the simply curious to highly organized, sophisticated, and formidable state and non-state enemies.

Citizens can, and do, change. A tinkerer can become a hacker; a cybercriminal could easily work for a nation-state; terrorists could start as disgruntled citizens.

Cyber actors cannot be put into single, static classifications. However, some categorization assists in the discussion:

- Placement: Insiders and outsiders: Are they part of the organization or not? Are the insiders disgruntled employees, financially motivated insiders, or merely those who unintentionally facilitate outside attacks?

- Organization: Loosely to highly organized: How coordinated are the activities?

- Motivation: Financial or political: Is there primarily a financial/economic motivation, or is there a political intent at play?

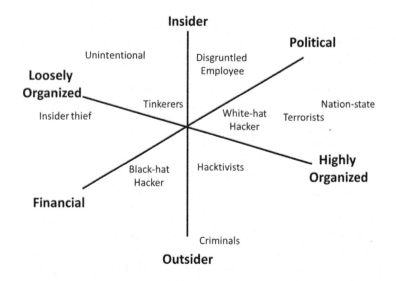

Motivation: Financial or Political

Financial and *Political* motivated cyber actors cross a wide spectrum and include hacktivists, criminals, terrorists, and adversary nation-states.

Financial

From low-level criminals to highly organized syndicates, the financially motivated actors are generally criminal. These are the *digital thieves*, who take advantage of the trust and lack of attribution found within the Internet. Their activities range from pilfering credit card data stealing health records, ransoming data, gaining access to bank accounts or stealing intellectual property with the intent of financial gain.

Political

Unlike the financially motivated, the politically motivated cyber citizen has a cause and wants to make a point. From changing the social order to creating awareness to improving national dignity, these actors fall into several categories: hacktivists, who want to make a political statement; terrorists, who seek to make a political statement through violent means; nation-states, who gather information for political or economic advantage or engage in acts of destruction in conjunction with other military operations.

Actions

This categorization of actions is based on analysis presented by MITRE Corporation in its November 2013 report, "Mapping the Cyber Terrain." Cyber actions are categorized in three ways: effect, extent, and intended impact.

Effects

As documented in many other works, cyber actors use a variety of methods with some specific *effects* in mind:

- Nuisance: Online harassment includes behaviors such as web defacement and cyber bullying

- Disruption: The targeting of a specific individual, organization, corporation, or government using technology that impedes said organization's ability to operate in the digital world, to include the use of botnets, spammers, and denial of service (DoS) / distributed denial of service (DDoS) attacks.

- Disinformation campaigns: Information campaigns, operations designed to filter the facts or shade the truth, to include the weaponization of memes, to influence not only individual behaviors but also entire societies.

- Criminal: Cybercrimes are the illegal attacks through cyberspace, as defined by the penal or administrative laws of a particular sovereignty. It covers all criminal offenses committed with the use of information technologies or communications networks.

- Espionage: A specific type of cybercrime, broadly defined as an operation to obtain unauthorized access to sensitive information by convert means. These attacks can target either intellectual property or sensitive information about the organization or government.

- Destructive: Destructive actions are those activities designed to intentionally destroy or long-term disable a designated target, such a virus that causes a device to overheat and burn up.

Extent

The *extent* of a cyber attack has two dimensions: scale and time frame. *Scale* is the operational reach of the cyber attacks. *Scale* answers the question of "How many targets?" *Time frame* is the duration of the attack. The time frame addresses the question of "How long does the attack last?" The extent component parallels the six criteria laid out by Michael N. Schmitt in his essay, "Computer Network Attack and the Use of Force in International Law: Thoughts on a Normative Framework," which is discussed in chapter 7, "Approach."

- Isolated: A single incident, generally focused on a specific organization or mission.

- Constrained: Brief, coordinated attacks directed to a specific organization or sector, bounded in scale and time.

- Persistent: Larger in time although not necessarily in scale, these attacks have a greater duration and are usually directed at highly specific targets to extract financial, mission-sensitive, or intellectual property–orientated data or to achieve a specific mission.

• Large-scale: Large in scale and in time, these attacks can be highly visible attacks that use virus or botnets, or they can be more deceptive, such as an advanced persistent threats (APTs), wherein code is inserted into an environment to reside for an extended period of time prior to either exfiltrating/removing data or disrupting a service.

Intended Impact

Intended impact can be identified in terms of the impact on services. The categories include:

• Negligible: An attacker can simply observe or monitor a service to gain intelligence about an organization, its mission, its people, or its strategy.

• Minor:

> • Restrict: Impede the ability of an organization to use a particular resource or service.

> • Alter: Change some feature or features of the resources providing the service, potentially impacting its operation, or changing data with the intent of altering the results of its usage.

• Significant: An attacker can terminate a service, possibly causing negative impacts on performance or safety.

CHAPTER 6

Cyber Axioms

Military and law enforcement theories and Sun Tzu have principles that apply to cyber conflict. As noted, the traditional military theories for land, sea, and air addressed four themes: identification, location, results, and execution. Law enforcement theory focuses on internal safeguards. Again, Sun Tzu has six underlying principles:

1. Know yourself

2. Know the enemy

3. Know the environment

4. Use all of your advantages

5. Exploit your enemy's weaknesses

6. Be deceptive and attack along unexpected lines.

An axiom is a statement upon which an abstractly defined strategy is based. To understand the cyber environment and prepare the organization, some cyber-specific axioms should be examined.

Network Effect–Driven

A *network effect* is defined as when a service or good gains value due to an increase in its utilization. The network effect has positively affected the marketing sector. By creating a platform for networks and software to reach a wide customer base. It has also raised the value of those networks and products.

As seen in the telecommunications industry, greater access to telephones meant that more people used them, and the value of the network to both the consumer and the corporations increased. This effect is also seen in software compatibility, in telecommunications standards, and in social networks.

The value and benefits of any given computer software comes about because of consumers and users who purchase the software or application. The software and application market is directly dependent on the network effect. As the number of application and software users increase, the value of those applications and software also increases. Social networks like Facebook and Instagram gain value as more people sign up and become users of these sites.

The network effect has also produced negative outcomes. One example is in the increased rate of cybercrimes. The expanding the number of people online has opened avenues for cybercriminals and other malicious actors. Large networks with a wide user base are highly susceptible to various security threats and breaches. Malicious actors use the connectivity and interoperability to identify and orchestrate cybercrimes like hacking via these networks.

Software with a high or ubiquitous user base also faces an increase in cyber threats. This can be seen through the proliferation of viruses. Viruses are malicious programs that are designed to alter the normal operations of computer software. The larger the value base of a piece of software, the higher the value for the creator of the virus due to the increased likelihood of finding a target. Software with many low-value users is not as lucrative a target as software that may have a limited number of users, but those users are conducting high-value operations.

Unrestricted

In the post–World War II, bipolar world, individuals, businesses, organizations, and governments have become accustomed to thinking that threats to a nation were from another state. However, the conflicts that defined the end of the twentieth century and the beginning of the twenty-first century have proven that belief wrong.

The principle of unrestricted warfare is based upon the 1999 book of the same name by Chinese colonels Qiao Liang and Wang Xiangsui. They wrote that war had changed and that nations, in an *unrestricted* fashion, should use "all means, including armed force or non-armed force, military and non-military, and lethal and nonlethal means to compel the enemy to accept one's interests." Conflicts would include attacks, without limits, on all elements of society. The front line was now within every facet of society

Cyber threats are no longer purely hypothetical in nature. In the hyper-connected world, cyber actors can attack any reachable element of another nation and society with one simple click.

No-Trust

One of initial fundamental design principles of the Internet, and in turn for cyber, is that actions and transactions are fundamentally assumed to be trustworthy.

This assumption holds even though it has been repeatedly proven false. This sounds dark and ominous. It is not. It is a neutral observation that no-trust, deception, and disinformation are at the heart of cyber conflict.

Robert Putnam, in *Bowling Alone*, put forward the convention of thick and thin trust. *Thick trust* is the trust embedded in personal relationships. These linkages are strong. As in the early days of the Internet, one would rely on these close-knit groups.

Thin trust is a form of social trust. It is the trust that extends beyond an individual's actual network. It extends into a sense of common networks and assumptions of eventual reciprocity. Thin trust is based more on community norms than personal experience.

The underlying premise and early cyber technologies were built for a thick-trust environment. Devices connect, exchange information, and then conduct a transaction.

In the face of an ever-widening expansion of sites and users, this model was found to be flawed. Over time, technology was forced to shift to a thin-trust model. Additional protections were designed and put in place.

Today, both assumptions have been proven faulty. Cyber is fundamentally a no-trust domain where deception is the coin of the realm.

No-trust is a security model and organizational philosophy, an approach that contradicts Levy's "Hacker Ethic." The architecture requires a no-trust default state for any and all entities. User, device, application, and data are assumed to be untrustworthy. No-trust segments the infrastructure to reduce the exposure of critical systems and to prevent lateral movement. No-trust requires access control throughout the environment and not simply at the boundaries. This is environmental-based protection where free, unfettered access is gone.

Omnidirectional

Cyber conflict comes from all directions, it is *omnidirectional*. Actions are conducted in different spaces at the same time. While organizational vulnerabilities and their mitigations are relatively static, the potential threats to which they correlate and assessing the defensive investment against the value of the assets potentially impacted may come in any direction.

Asymmetric

Cyber conflict is *asymmetric* in nature, having a condition where the power relationship between the various actors differs, usually significantly.

Variable Barriers to Entry

Generally, barriers to entry are the costs or obstacles that prevent competitors from entering a field. Depending upon the environment, actor, and actions, the *barriers to entry* for cyber conflict vary, ranging from exceptionally low to exceptionally high.

It does not cost much to conduct low-level cyber conflict. Cloud-based servers and almost universal Internet access allows actors to rapidly gain both the resources and connectivity necessary to conduct operations. Given the highly opportunistic nature of the threat and the large number of potential attack points, the cost to defend is potentially high. In a highly connected world, cyber defenders are required to defend everything, everywhere, equally.

Conversely, with planning, an understanding of the environment, and an asset value assessment, organizations can prepare and allocate defensive resources more heavily to the assets they most highly value. Organizations can take defensive actions to reduce their risk. The regular "rebaselining" of their infrastructures, routinely destroying and regenerating assets within their environment, increases the actual or opportunity cost to attackers, and, potentially, causes them to go elsewhere.

Atemporal

The nature of time and distance has changed. With a virtual and physical nature, the *atemporal* nature of cyber conflicts means that actions occur without relationship to time; while adhering to the laws of physics, cyber actions can occur in milliseconds across the globe. Concurrently, actions can also be programmed to automatically happen at a different times and places; these actions can be dependent or independent of other activities. This environment exists without the familiar time and space dimensions.

Zero-Defect

In the *zero-defect* environment of cyber conflict, organizations can no longer make mistakes. In cyber conflict, malicious actors exploit opponents' weakness and mistakes. In a technology-based environment, those weaknesses often include misconfigurations or out-of-date software.

Deceptive

Deception is commonly defined "as an interaction between two parties, where one party wants the second party to accept as true something that is not." In cyber, deception can readily happen, and it is often the reason that attackers are successful.

While in the physical world, there are routinely cues to establish the veracity of a person, of information, or of an assertion, the cyber environment does not readily provide those same cues. The cyber environment does not have the same face-to-face interactions found within the physical world.

Cyber is without body language, voice tone, or facial expression. Electronic exchanges do not provide the necessary behavioral cues. Without those cues, it is easy to be tricked.

Without a physical presence, it also is hard to determine the quality of a product, since the product itself cannot be examined.

In cyber, deception is rampant. Through impersonation, the use of e-mails to gain passwords, and links to false websites where a user enters personal data, attackers gain access to information and resources that they are not authorized to have.

Beyond deceiving people, attackers also mislead computers. Using faked information on the data transmitted between machines, attackers can make one machine look like another. Once an attacker gains access to the system, an organization's systems can be duped into obeying the attacker's commands.

Attribution

Attribution is "the action of regarding something as being caused by a person or thing." Given the virtual and deceptive natures of the cyber environment, attribution is difficult. E-mail addresses and websites that appear authoritative can be rapidly established and just as rapidly taken down. Information regarding source and intent can be quickly and easily changed. Actors can have computers impersonate another, using these systems as camouflage or as a launching point for an attack.

Attribution is difficult because of the Internet's fundamental design. The communications protocols that allow systems to "talk" to each other and for network traffic to flow are challenging to trace. With layers of intermediate machines between attackers and targets, tracing a malicious action back to the source can be problematic. Further, often these machines are located throughout the world, raising the issue of cyber sovereignty.

Disingenuous

Disinformation is the use of intentionally false or misleading information used in way to misdirect or deceive a target audience. Its aim is to mislead an enemy.

Distorted and outright false information on social media confuses public perceptions, sowing doubt and division. Using half-truths and biases, disinformation undermines the official version of events, to the point of fostering the notion that there is even a true version.

This disingenuity serves to confuse and distract organizations. They freeze, unable to focus or make decisions. Given the virtual and deceptive natures of the cyber environment, attribution becomes difficult.

Rhizomatic

The malicious cyber actor's organizational structure is very much a "rhizomatic" command system. With a nominal, visible, hierarchical control system above ground, the true control system resides hidden, below ground. According to British general Rupert Smith, as with weeds, a *rhizomatic* command structure is "is a horizontal system, with many discrete groups. It develops to suit its surroundings and purpose in a process of natural selection, and with no predetermined operational structure; its foundation is that of the social structure of its locale."

CHAPTER 7

Approach

There are a multitude of frameworks for detecting and resolving the problems associated with cyber conflict. These frameworks have some underlying themes:

- How to recognize the problem

- How to remediate the problem

- How to respond to the problem

Recognize
"Contextual Awareness"

The cyber environment is constantly shifting. In response to new business drivers, consumer needs, and technologies, the components and configurations change over time. New mobile devices, applications, or network devices constantly and consistently create or remove vulnerabilities. Attackers respond and adapt to these changes. Traditional approaches of siloed monitoring tools and isolated static analyses are ineffective for the rapidly evolving attack/defense situations.

Context and contextual awareness are fundamental. As previously published,

> In this "Internet of everything" world, information technology is found within a highly integrated and inter-connected grouping of networks, systems, and applications. Within this environment, there is always the potential for gaps that allow compromise.

Contextual awareness is the understanding of the environment through a series of "cyber planes." These planes are technology, components, and information. The technology plane are those leading-edge research and technologies that provide the nation with a decisive edge in mission delivery. Attention is on keeping the secrets in. The second category are the components. Components are those mission critical hardware and software elements, generally sourced through a commercial supply chain. In this case, the emphasis is on keeping bad things out. The third and final part of this triad is the information itself. In this instance, the focus in on ensuring that organizations are able to keep critical information from getting out.

In this hyper-connected world, these three planes are then overlaid by three other concepts: how the systems are being constructed; how the systems are being delivered, and where the systems are being delivered to. Systems providence means that organizations know who built what. Where systems are being delivered from has become more complicated in the world of software as a service, data center consolidation, and cloud computing. Finally, where information is delivered matters. There are different cybersecurity needs for warriors in a deployed environment than for team members working on a budget submission. (Sienkiewicz, 2013)

With a structured process, contextual awareness and the detection of anomalous behavior can be achieved. This process should include five elements:

- Risk manage: Using a risk-management framework, establish security policy and controls that are enforced and validated through an audit process. A risk-management framework allows the organizations to know themselves, their vulnerabilities, what are the threats that could act on their vulnerabilities, what is the likelihood of the threats being realized, and what is the value of the asset(s) being defended. Multiple standards bodies have put forward risk-management frameworks for their respective industries.

• Baseline: Collect data using a baseline to understand the technology configuration and environment, which allows for the identification of technical vulnerabilities and organizational weaknesses.

• Assess: Using vulnerability assessment, validate the state of the organization's technology systems.

• Monitor: Through real-time observation and tracking, provide the organization's risk status to the business operations team, the cyber professionals, and the executive leadership.

• Correlate: Through correlation, analyze and prioritize the available data to more fully understand the operational environment and associated risks.

Risk Manage

Risk management is a continual process to ensure cyber safety, not a simple audit to check a compliancy box. An organization's risk-management framework is its unified information security framework. Risk management resides with the organization's most senior leadership. Organizational leadership is responsible for monitoring performance, to include requiring the updating of security controls and remediating any findings. The risk-management process should:

1. Identify a clear management "owner" of cybersecurity risk;

2. Decide where in the organizational structure primary oversight responsibility will lie;

3. Require the designated cybersecurity owner to communicate with the stakeholders using business risk language that they understand; and

4. Integrate the baseline, assessment, monitoring mechanisms, correlation, remediation, and response processes into a contextually aware depiction of the organization's cyber environment, including its most valuable assets and vulnerabilities.

Baseline

A *baseline* is "a conceptual blueprint that defines the structure and operation of an organization," while making visible the organizational seams and gaps. These constituents are part of a highly interrelated set of relationships with every other element. Technologies, partners, and other attributes are part of the symbiotic nature of the cyber environment, an environment that varies in scale from micro level to the enterprise level. This is also known as an *enterprise architecture*.

The baselining process discovers and classifies devices, correlates the devices, models vulnerabilities, analyzes the result, and provides a product that lets an organization prepare for attacks, prioritize investments, and manage vulnerability risks. The goal of the baseline process is to know what the organization has and how it connects, and to allow for the development of the normal and abnormal models.

The baseline uses outputs from vulnerability scanners and firewalls and includes third-party cyber threat information to depict potential attacks routes against the enterprise. These models represent the possible cyber attack paths. The models permit the development of what-if scenarios and the establishment of the optimal mitigation methods.

Baselining provides a foundation for modeling and analysis that allows an understanding of vulnerabilities.

Assess

Assessment—specifically vulnerability assessment—is the identification and quantification of security vulnerabilities in an environment. Assessment identifies risk within the environment using information technology security to test performance and reliability within a corporation or the government reduction of the underlying weakness to a tolerable level.

Assessments show which vulnerabilities are present. However, without an awareness of the cyber environment, assessments cannot differentiate between flaws that can be fully exploited to cause damage and those that cannot be. It is through the contextual awareness provided by an assessment against the baseline and a risk-management framework that organizations can understand which flaws pose a meaningful threat to their environment and reduce the underlying weakness to a tolerable level.

As the American security technologist Bruce Schneier wrote, "Defending computer system often requires people who can think like hackers." Because of the need to have objectivity and to reduce the potential of insider threats, it is important to have a third party with no association with the organization to do penetration testing. Third-party vulnerability testing helps an organization to see how an outside hacker will identify a system weakness.

Similar to a financial audit, an organization should uses a structured independent and objective vulnerability assessment and penetration-testing approach to develop a detailed view of the threats that faces its data and. This enables the organization or the corporation to prioritize resource allocation to prepare to protect its data.

The cyber community should adopt the same language as the Institute of Internal Auditors (IIA) Standard Board (IASB). The IASB states that "Independence is the freedom from conditions that threaten the ability of the internal audit activity or the chief audit executive to carry out internal audit responsibilities in an unbiased manner." Additionally, the IASB defines objectivity as "an unbiased mental attitude that allows internal auditors to perform engagements in such a manner that they believe in their work product and that no quality compromises are made. Objectivity requires that internal auditors do not subordinate their judgment on audit matters to others."

Monitor

To effectively assess the baseline, organizations need to see what is going on within their environment. Network and systems *monitoring* is the way that an organization begins to comprehend itself.

In this complex cyber environment, the network effect comes to bear. Each machine is dependent upon the other machines. Susceptibility to attacks hinges upon the vulnerabilities of the other machines in the network.

Network- and systems-monitoring tools generate a great, great deal of data and corresponding alerts. These data and alerts are normally without context. Often undefined and ambiguous, they may be of little consequence or even incorrect, or they may be critically important.

Correlate

Within the context of their cyber environment, organizations must understand not just the vulnerabilities of each device; *correlation* allows organizations to understand the connections and dependences. Correlation allows for this understanding.

Using the correlation model put forward by the RAND Corporation in its 2004 study, "Out of the Ordinary," it is clear that deciding which vulnerabilities could trigger the greatest loss is not straightforward. Data losses can have long-term consequences up to and including the viability of the company because of the financial costs of remediation, loss of customers and reputation, and legal liability. The loss of key intellectual property could threaten the organization's value proposition. Loss of credit card data has caused customers to become wary. The loss of sensitive data might jeopardize the organization's ability to run its business.

Combining distinct bits of data to understand threats is a difficult task. The process of analyzing involves going through huge volumes of data. It also requires exceptional attention to detail and potentially an investment to be able to focus on the important findings.

This is not only a question of connecting the dots; it is also a question of seeing abnormal dots and, in turn, preventing them from occurring. These dots would be the difference between normal and suspicious behavior. The baseline normal behavior would be identified and categorized, and any anomalies or deviations from defined normal behavior would be prioritized for investigation and remediation.

There exist few direct indications as to what data is significant and how to connect the data together. A common strategy is to prioritize data by establishing of patterns of similar previous events.

In this approach, the first step is to baseline the environment, documenting what is considered normal. The second step is to abstract the data within the pattern of metrics identified for observation.

The third step is to associate the information using automated relationship agents to determine relationships between the existing and the new dots. Fourth, hypothesis proxies create possible interpretations for the connected dots. They also create consistent testing strategies to identify whether the hypotheses are accurate. Fifth, the outcome of the processes are ranks the results, with the high risks is passed on to the remediation team. The RAND Corporation model asserts that like human problem solvers, the correlation schema allows iterative, dynamically bespoken analysis.

Remediate
"Industrialized" Remediation

Once an organization recognizes a vulnerability, it needs to act to prevent it from being used by cyber attackers. Remediation is the process of taking those actions through patching, software upgrades, installation of new equipment, and/or the reconfiguration of existing equipment.

The proliferation of computers, network devices, mobile devices, and software requires organizations to ensure that all these elements are current, updated, and secured to mitigate the risk. A software patch is software designed to update, improve, fix, or support the functionality of a program by fixing some forms of security vulnerabilities, bugs, stability, user interfaces, and other elements of software (Lindstrom, 2004). Automated software-configuration management and patching programs identify vulnerabilities and resolve potential issues before unauthorized actors can gain access. The remediation process is industrialized.

A process becomes industrialized when it transitions from being the providence of a craftsman to a repeatable manufacturing function. The technology industry has been slow to adopt industrialized operations, of which remediation is a part. This in due to a perceived notion that individual product lines and business units had unique support and technology needs.

Manual patching processes are susceptible to errors introduced by the involvement of the human factor in the process; done properly, the automated patching process is less susceptible to errors or mistakes. This includes the requirement to have a rigorous testing and validation process and a mechanism in place to automatically remove, roll-back, patches. Smith et al. (2015) observed that the use of automated patching systems reduces the risk of errors in patching and eliminates the likelihood of making mistakes moving forward. Qi et al. (2014) argued that the use of automated patching ensures that the system automatically searches the programs and software and patches any vulnerabilities and/or bugs before they are deployed should be implemented. This industrialization of remediation reduces mistakes or errors in the patching process, and the need for the organization's personnel to engage in rework activities aimed at rectifying or remedying the mistakes and or mistakes that could result from a manual patching process. A caveat is that in complex systems, an automated process can lack awareness of the various dependencies in the subsystems.

Respond
"Active Discouragement"

Every cyber actor has a weakness. While malicious actors have profited by organizational weaknesses, weaknesses can be exploited by all parties.

Active discouragement is the systematic process of deterrence including the inclusion of preemptive measures. It has a goal of not only discouraging malicious behavior, but also intentionally and legally inflicting a cost upon the adversary. Active discouragement directs an organization's response toward taking advantage of the opposing actors' weaknesses.

The question posed at the beginning of this book was, "How do I keep my 'stuff' safe?" The traditional responses have been geared at defensive remediation actions, to include the creation of internal policies, keeping computers updated, conducting awareness training, frequently changing passwords, and hiring security experts.

Active discouragement could be installing obfuscation technologies that disguise the organization's internal network or data traffic. It could be the installation of intrusion systems that require attackers to take longer to perform their actions. Organizations could create poisoned wells of data that make attackers unsure of what is valid and what is invalid, thus destroying the target's potential value. Systems administrators could embed malware within stored files, files that can corrupt the attackers' systems or start processes that automatically consume all of the available resources, essentially creating an internal denial-of-service attack. Tracing technologies can be deployed, allowing for organizations to track the location of all files. Encryption could become widespread. Individuals and companies could take legal action against network providers who allow attackers or malware to gain access to their systems.

Individuals

As citizens with agency and due to the network effect, individuals have an increased amount of responsibility to safeguard themselves and their respective organizations in the digital world. The greater the number of secured machines, the more secure the overall environment. Some individual measures of active discouragement include:

- Protect:

    - Password: The single biggest point of failure remains the user's password. Individuals should not only use a strong complex password, passwords should also be regularly changed. Further, the same password should never be used across multiple public services where the password from one compromised account can then be used for unauthorized access on another service.

• Multifactor authentication: Multifactor authentication is the use of more than one method of authentication, such as a text message to a mobile device when a password is changed or when a service is accessed by a new device.

• Proper use of social media: When using social media, individuals should avoid disclosing much personal identity to ensure that attackers cannot easily craft cyber attacks directed specifically at them, such as a spear phishing.

• Backup: Individuals should always have backup data to prevent data loss and to minimize the impact of any attempt of a cyber attacker to hold their data hostage, a process commonly known as *ransomware*.

• Educate: It is an individual's responsibility to educate himself or herself, friends, and family of the importance of cybersecurity and how to reduce cyber threats.

• Maintain: In regard to cyber maintenance, the network effect prevails. In short, the better one machine is in terms of patch and vulnerability management, the more likely it is to ward off infection and thus will not infect others. The greater the number of protected machines, the securer the entire environment. Conversely, the greater the number of unprotected machines, the more potential attack vectors. Individuals need to keep their applications and operating systems current.

• Encrypt: Encryption is the algorithmic scrambling of data, data either stored on a computer system (also called data at rest) or in transit (also called data in motion), to prevent unauthorized access. Most operating systems come with encryption technology that simply needs to be turned on. Many browsers, and virtually all websites, have encryption features.

Organizations

Every organization has roles to play, both for their own well-being and to protect customers' information from cyber attacks. Senior management is directly responsible for the organization's cybersecurity posture. Some of the following measures should be taken by organizations to ensure cybersecurity:

- Train: Corporate organizations should bring an awareness to their workforce about the need for cybersecurity. They should educate every member of the organization regarding cyber threats and how to identify them. They should require at least annual cyber awareness training.

- Assess: There should be a periodic assessment of the organizations approach to cybersecurity. The assessment needs to include not just the identification of assets, location of assets, the classification the assets' value, and threat models; it needs to ensure that the approach aligns with the organization's strategy, that senior-level decision makers are fully informed, the resources have been allocated, and that the entire team has been trained.

- Monitor: Monitoring—specifically continuous monitoring—is the ongoing surveillance of what is happening on the organization's networks and systems. It is an integrated approach, combining skills and services to recognize and remediate cybersecurity incidents and suspicious activities.

• Protect: Through encryption and computer security protection devices, they should safeguard the company's and customers' data electronically so that unauthorized actors cannot gain access to the data, nor can authorized actors use data inappropriately.

• Mandate: Through their collective purchasing power, organizations can force vendors to adhere to standards and maintain cyber standards. Vendors need to be liable for flaws. Organizations need to require telecom and Internet service providers to verify domain providers. Contractual agreements need to remove the exception from liability clause for the transmission of viruses and malware.

• Ally: In chapter 8, subchapter 7, "Maneuvers," *The Art of War* addresses the necessity to build alliances and form partnerships. The financial and law enforcement communities have realized that alliances and partnerships work. Interpol and the Society for Worldwide Interbank Financial Telecommunication (SWIFT) are two examples of international cooperation. For the law enforcement and financial communities, respectively, these two organizations are pursing recourse in the area of cyber conflict as they develop industry-wide standards and best practices. These alliances provide a model for others, demonstrating that organizations can initially share information and eventually grow to the adoption of common procedures and technology standards. The eventual sharing of cyber threat information is essential. Through alliances, partners receive information that they may not otherwise get. As the alliances grow, they can develop standardized approaches to security assessment, authorization, and collective continuous monitoring.

Governments

Indirect Responses

Governments have a responsibility to protect the nation and its people. This includes a role in cybersecurity. Governments need to work alongside industries to keep the infrastructure secured. In a public-private partnership, governments need to set out measures to fight against cyber attacks and threats. The measures should include the following:

- Mandate:

    - Prosecution of organizations that enable the conduct of malicious cyber activities, including the network providers, software developers, and the financial community.

    - Providers to notify customers whose machines have been infected by a botnet.

    - Standards that limit the ability of organization to connect critical infrastructure industrial control systems to the public network.

    - Disclosure of cyber risk factors in the filing for all publicly traded companies.

    - Public audit standards to include a positive requirement for cybersecurity.

- Educate:

    - Establish and publish cybersecurity best-practice guides to provide individuals and organizations with a foundation to build their internal practices.

    - Establish cybersecurity curriculum to have youths become familiar with cybersecurity and how to reduce the risk of cyber threats.

- Allow: Organizations to exchange threat information without fear of prosecution for collusion.

- Prohibit:

    - Government organizations from using services that do not include embedded security.

    - Government organizations from doing business with any provider that is known for hosting botnets.

Direct Response

As noted earlier, technologists, businesspeople, and the military commonly share one attribute—they like to get things done. They have a strong preference to action.

In the face of cyber attack, they want to respond; they want to figure out how to strike back.

Michael N. Schmitt's Analytical Framework

Severity

What is the scope and intensity of an attack? The larger the amount of damage, the more likely the cyber attack should be considered an armed attack.

Immediacy

What is the duration of a cyber attack? The longer the duration of the attack, the more likely it should be considered an armed attack.

Directness

Does it directly cause harm? If the attack directly causes harm, then it is more likely considered to be an armed attack.

Invasiveness

Does it cross sovereign borders? The more invasive the cyber attack, the more it looks like an armed attack.

Measurability

Can the damage be measured? The more the attack can be measured, the more likely it is to be considered an armed attack.

Presumptive legitimacy

Does the attack conform to acceptable norms? If the action conforms to conventional behavior, then it may be considered legitimate.

Direct response is difficult. Direct response requires attribution. Further, at least under American law, direct response is reserved for the military, the so-called Title 10 organizations.

There are two major issues surrounding direct response. First, attribution is exceptionally difficult. Second, the determination of when a cyber attack constitutes an armed attack.

The attribution requirement prevents a direct response because of the extreme difficulty of attributing a cyber attack during the actual attack. Although at times states can trace cyber attacks back to computer servers in another state, conclusively ascertaining the identity of the attacker requires intensive, time-consuming investigation normally with the assistance of the state of origin.

Cyber attacks come in many different forms. While it may seem natural to categorize these attacks as armed attacks, the international legal community has been reluctant to adopt explicit standards.

In regard to the criteria of when a cyber attack constitutes an armed attack, Michael N. Schmitt laid out six criteria in his essay "Computer Network Attack and the Use of Force in International Law: Thoughts on a Normative Framework" for determining if a cyber attack was an armed attack. These criteria are *severity*, *immediacy*, *directness*, *invasiveness*, *measurability*, and *presumptive legitimacy*.

In the absence of direct military force but in the face of constant cyber conflict, how should individuals, businesses, organizations, and governments think about their response to cyber conflict? *The Art of War* provides an approach.

CHAPTER 8

The Art of War

Background Notes

The translation used in this text is the Samuel B. Griffith's 1963 translation. A United States Marine, Brigadier General Griffith (1906–1983) was a well-known historian and lecturer. He was stationed in China during the last days of the Chinese Nationalist rule in mainland China.

The Art of War offers a basic framework for understanding conflict, comparing strengths and weaknesses, and provides a basic framework for achieving success. The first two subchapters focus on planning and the cost of conflict. The middle subchapters concentrate on understanding tactics, disposition, force, weaknesses, and strengthens. The penultimate section of the book discusses responses. The final subchapter stresses the need to collect and analyze information.

In this work, each subchapter begins with "The Stage," a short summary of the text and applicability to cyber conflict. Next, the Griffith's translation of *The Art of War* is provided. Finally, corollaries focused on cyber conflict with specific textual references are presented.

I: Preparation

The Stage

This subchapter serves as both an introduction to *The Art of War* and as an overview for *The Art of War*. The title has many translations, including "Estimates," and "Planning." However, "Preparation" was chosen because it is the one that most directly applies to cyber conflict.

In "Preparation," Sun Tzu outlines an approach to holistically understand the environment through an analysis of mission, climate, ground, command, and methods. He describes how to evaluate strengths and weaknesses, an evaluation that can apply both to the organization and potential adversaries. This chapter concludes with the idea that information is limited and routinely incomplete.

As *The Art of War* clearly implies, organizations should respond to conflict not with checklists but with planning, awareness, and preparation. The same idea holds true for cyber conflict. Organizations need to plan to recognize, remediate, and respond to cyber conflict.

The cyber preparation process seeks to identify and prevent the exploitation of weaknesses, while it encompasses people, processes, and technology. Preparing for a cyber response is about managing security with a multilayered approach and is focused on ensuring that the real objective of safeguarding data and continuous operations in the face of cyber attacks is attained.

SUN TZU said

1. War is a matter of vital importance to the State; the province of life or death; the road to survival or ruin. It is mandatory that it be thoroughly studied.

2. Therefore, appraise it in terms of the five fundamental factors and make comparison of the seven elements later named. So you may assess its essentials.

3. The first of these factors is moral influence; the second, weather; the third, terrain; the fourth, command; and the fifth, doctrine.

4. By *moral influence* I mean that which causes the people to be in harmony with their leaders, so that they will accompany them in life and unto death without fear of moral period.

5. By *weather* I mean the interaction of natural forces; the effects of winter's cold and summer's heat and the conduct of military operations in accordance with the seasons.

6. By *terrain* I mean distances, whether the ground is traversed with ease or difficulty, whether it is open or constricted, and the chances of life or death.

7. By *command* I mean the general's qualities of wisdom, sincerity, humanity, courage, and strictness.

8. By *doctrine* I mean organization, control, assignment of appropriate ranks to officers, regulation of supply routes, and the provision of principal items used by the army.

9. There is no general who has not heard of these five matters. Those who master them win; those who do not are defeated.

10. Therefore in laying plans compare the following elements, appraising them with the utmost care.

11. If you say which ruler possesses moral influence, which commander is the more able, which army obtains the advantages of nature the terrain, in which regulations and instructions are carried out, which troops are the stronger;

12. Which has the better trained officers and men;

13. And which administers rewards and punishments in a more enlightened manner;

14. I will be able to forecast which side will be victorious and which defeated.

15. If a general who heeds my strategy is employed he is certain to win. Retain him! When one who refuses to listen to my strategy is employed, he is certain to be defeated. Dismiss him!

16. Having paid heed to the advantages of my plans, the general must create situations which will contribute to their accomplishment. By *situations* I mean that he should act expediently in accordance with what is advantageous and so control the balance.

17. All warfare is based upon deception.

18. Therefore, when capable, feign incapacity; when active, inactivity.

19. When near, make it appear that you are far away; when far away, that you are near.

20. Offer the enemy a bait to lure him; feign disorder and strike him.

21. When he concentrates, prepare against him; where he is strong, avoid him.

22. Anger his general and confuse him.

23. Pretend inferiority and encourage his arrogance.

24. Keep him under a strain and wear him down.

25. When he is united, divide him.

26. Attack where he is unprepared; sally out when he does not expect you.

27. These are the strategist's key to victory. It is not possible to discuss them beforehand.

28. Now if the estimates made in the temple before hostilities indicate victory it is because calculations show one's strength to be superior to that of his enemy; if they indicate defeat, it is because calculations show that one is inferior. With many calculations, one can win; with few one cannot. How much less chance of victory has one who makes one at all! By this means, I examine the situation and the outcome will be clearly apparent.

Cyber Corollaries

1. War is a matter of vital importance to the State; the province of life or death; the road to survival or ruin. It is mandatory that it be thoroughly studied.

If there is only one thing to learn from *The Art of War*, it is the necessity to plan and prepare. Success, in warfare or in cyber conflict, is only achieved through thoroughly studying the organization, the opponents, the environment, and the objectives. In this case, *the State* refers to both public and private organizations.

For cyber preparation, Sun Tzu's advice of "knowing yourself" and "knowing the enemy" sets the stage. Organizations need to have a full awareness of their environment and to prepare for the contingencies that will occur.

With the constant threat of a cyber attack and routine cyber breaches, organizations need to be prepared to respond to the inevitable. The direct responsibility of organizational leadership, preparation includes a baseline that allows the organization to see itself from internal and external perspectives; an asset valuation to establish what information is valuable and the steps necessary to safeguard it; a cyber incident response plan that is current, accessible to decision makers, and contains specific activities.

5. By weather, I mean the interaction of natural forces; the effects of winter's cold and summer's heat and the conduct of military operations in accordance with the seasons.

Sun Tzu points out that commanders have no control over the weather and must adjust their operations "in accordance with the season." In a similar manner, organizations have no control over their entire cyber environment. They can control pieces of it; they can influence other aspects. They can endeavor to forecast and create guides for cyber and risk-management activities, although these only serve as plans that will be adapted in the face of cyber attacks.

12.   *Which has the better trained officers and men;*

Even prior to one of the most famous lines in *The Art of War*, the line regarding deception, Sun Tzu discusses the need to train the officers and men. With the ever-increasing number of cyber attacks, it is critical to have enough professionals ready to respond to events that happen at the "speed of cyber."

With a multitude of information sources and an abundance of tools, cyber professionals need training, having tools is not enough. Cybersecurity effectiveness can be increased through standardized, focused, resource-efficient, role-appropriate training, with specific skills enhancement as part of a comprehensive workforce training plan, including a hands-on component with recurring individual and collective tasks. This standardized training should provide the organization with more effective professionals who can operate and defend the organizations networks and systems. In short, cyber professionals need to be fully trained and certified to common individual and team standards that would enable a successful cyber defense.

Cyber training is part of business operations. It should not considered a standalone task or solely a requirement for cyber professionals.

The often-ignored issue is that the rest of the workforce, the people who are routinely the targets of cyber attacks, needs to be trained. Cyber awareness and an understanding of cyber best practices need to be as ubiquitous as using a browser, a smartphone, or a word-processing application.

17.   *All warfare is based upon deception.*

Quoted countless times, for many casual readers of *The Art of War*, this line is one of the best known. Cyber conflicts have proven Sun Tzu prescient, since deception is an integral part of the cyber conflict. Through the inserting of false data into sensors, the jamming of communications, or social engineering, cyber conflicts are rife with impersonations and falsehoods. Attackers deceive people; attackers deceive computers; attackers deceive the sensors used to detect them.

The anonymous nature of the cyber environment allows attackers to act almost without consequences. Yet, as they act they leave, they leave behind an electronic trail of their activities, their digital footprints. These digital footprints often expose their personal, technical, or organizational information. By tracing back these footprints and understanding the paths that the attackers have taken, organizations can use the information to prevent corporate espionage, competitive intelligence, or cyber attacks.

II: Conflict's Cost

The Stage

Often translated as "Going to War," this subchapter describes the economic, political, and military cost war imposes upon all parties. For Sun Tzu, protracted campaigns stretch the nation's resources. This is true not just for the attacked but also the attacker.

The Art of War clearly states that all war is costly. Cyber conflict, in particular, is expensive. There is an economic cost of goods directed toward the conflict or never produced. There is an opportunity cost of endeavors not undertaken due to cyber risks. There is a direct financial cost when organizations upgrade equipment.

The friction induced by the need to safeguard imposes a cost. Data exchange is no longer done by a simple electronic handshake; rather, it requires an expensive filtering and sterilization process. There is a cost when boards of directors and senior leaders need to spend time to understand the cyber problem, to develop solutions, and to monitor compliance rather than focusing on business or organizational needs. There is a cost when individuals cannot automatically trust that their online transactions or interactions are either safe or private.

Resources are expended to recognize; resources are expended to remediate; resources are expended to respond. The attention and effort required to safeguard the environment diverts organizations from their primary focus or mission.

SUN TZU said

1. Generally, operations of war require one thousand fast four- horse chariots, one thousand four-horse wagons covered in leather, and one hundred thousand mailed troops.

2. When provisions are transported for a thousand *li* expenditures at home and in the field, stipends for the entertainment of advisers and visitors, the cost of materials such as glue and lacquer, and of chariots and armor, will amount to one thousand pieces of gold a day. After this money is in hand, one hundred thousand troops may be raised. (Author's note: A *li* is a traditional Chinese measurement of distance usually about a third of an English mile)

3. Victory is the main object in war. If this is long delayed, weapons are blunted and morale depressed. When troops attack cities, their strength will be exhausted.

4. When the army engages in protracted campaigns the resources of the state will not suffice.

5. When your weapons are dulled and ardor damped, your strength exhausted and treasure spent, neighboring rulers will take advantage of your distress to act. And even though you have wise counselors, none will be able to lay plans for the future.

6. Thus, while we have heard of blundering swiftness in war, we have not yet seen a clever operation that was prolonged.

7. For there has never been a protracted war from which a country has benefited.

8. Thus those unable to understand the dangers inherent in employing troops are equally unable to understand the advantageous ways of doing so.

9. Those adept in waging war do not require a second levy of conscripts nor more than one provisioning.

10. They carry equipment from the homeland; they rely for provisions on the enemy. Thus the army is plentifully provided with food.

11. When a country is impoverished by military operations it is due to distant transportation; carriage of supplies for great distances renders the people destitute.

12. Where the army is, prices are high; when prices rise the wealth of the people is exhausted. When wealth is exhausted the peasantry will be afflicted with urgent exactions.

13. With strength thus depleted and wealth consumed the households in the central plains will be utterly impoverished and seven-tenths of their wealth dissipated.

14. As to government expenditures, those due to broken-down chariots, worn-out horses, armor and helmets, arrows and crossbows, lances, hand and body shields, draft animals and supply wagons will amount to sixty per cent of the total

15. Hence, the wise general sees to it that this troops feed on the enemy, for one bushel of the enemy's provisions is equivalent to twenty of his; one hundredweight of enemy fodder to twenty hundredweight of his.

16. The reason troops slay the enemy is because they are enraged.

17. They take booty from the enemy because they desire wealth.

18. Therefore, when in chariot fighting more than ten chariots are captured, reward those who take the first. Replace the enemy's flags and banners with your own, mix the captured chariots with yours, and mount them.

19. Treat the captives well, and care for them.

20. This is called, "winning a battle and becoming stronger."

21. Hence what is essential in war is victory and not prolonged operations. And therefore the general who understands war is the Minister of the people's fate and arbiter of the nations' destiny.

Cyber Corollaries

3.   Victory is the main object in war. If this is long delayed, weapons are blunted and morale depressed. When troops attack cities, their strength will be exhausted.

and

4.   When the army engages in protracted campaigns, the resources of the state will not suffice.

As Sun Tzu clearly points out, fighting takes resources. Resources are limited. When depleted, the nation is left defenseless. In short, continual warfare will destroy a country.

In the same vein, continual cyber conflict diverts individuals, organizations, corporations, and governments from their primary missions. Cyber conflict, and in turn the activities necessary to provide security, is expensive. These scarce resources could be used elsewhere.

The cost of cyber conflict is not simply financial or the loss of intellectual property. As noted, there are other opportunity costs when a company cannot focus on its core mission. There is the cost required to repair brand damage. There is the cost of cleanup.

9.   Those adept in waging war do not require a second levy of conscripts nor more than one provisioning.

In the zero-defect world of cyber, cyber actors need to install and operate the components of their environment right the first time, and with remediation, to do it right on a continual basis. With proper planning and understanding, organizations can allocate resources to defend their highest value assets from the beginning. Further, while it may initially take more time and resources, the cost to build security from the onset is significantly less than the cost to retrofit the infrastructure.

III: Tactics

The Stage

Often translated as "Attacks," "Strategies," "Offensive Strategies," or "Planning the Attack," *Tactics* describes the nature of strength and what is required for an army to successfully move into new territories. *Tactics* refers to the process of planning, coordinating, and giving general directions to operations with the aim of meeting the overall military, political, and/or economic objectives. Tactics implement strategies through the short-term decisions made about the employment and movement of resources and how those resources are used in the operation. According to the theorist Carl von Clausewitz, "Tactics is the art of using soldiers in a battle, while strategy is the art of using battles to win."

Sun Tzu specifically addresses where to focus an organization's efforts to gain maximum advantage at the minimal cost. He counsels to first attack the enemy's strategy, then at the alliances, then army—and if those attacks have failed, then attack the cities. He builds on the notion of knowing when to attack and to fight only when victory is assured.

The Art of War repeatedly addresses attacking weaknesses. In the context of conflict, cyber offensive tactics are somewhat like traditional tactics, specifically in attacking weaknesses and attacking asymmetrically.

For most individuals, organizations, and agencies, cyber tactics are more defensive in nature. Cyber defensive tactics significantly differ concentrating on the protection of the data and not particular devices. Further, legal constraints, attribution requirements, and sovereignty issues generally preclude direct action.

SUN TZU said

1. Generally in war the best policy is to take a state intact; to ruin it is inferior to this.

2. To capture the enemy's army is better than to destroy it; to take intact a battalion, a company, or a five-man squad is better than to destroy them.

3. For to win one hundred victories in one hundred battles is not the acme of skill. To subdue the enemy without fighting is the acme of skill.

4. Thus, what is of supreme importance in war is to attack the enemy's strategy.

5. Next best is to disrupt his alliances.

6. The next best is to attack his army.

7. The worst policy is to attack cities. Attack cities only when there is no alternative.

8. To prepare the shielded wagons and make ready the necessary arms and equipment requires at least three months; to pile up earthen ramps against the walls an additional three months will be needed.

9. If the general is unable to control his impatience and orders his troops to swarm up the wall like ants, one-third of them will be killed without taking the city. Such is the calamity of these attacks.

10. Thus, those skilled in war subdue the enemy's army without battle. They capture his cities without assaulting them and overthrow this state without protracted operations.

11. Your aim must be to take All-under-Heaven intact. Thus your troops are not worn out and your gains will be complete. This is the art of offensive strategy.

12. Consequently, the art of using troops is this: When ten to the enemy's one, surround him;

13. When five times his strength, attack him;

14. If double his strength, divide him.

15. If equally matched, you may engage him.

16. If weaker numerically, be capable of withdrawing;

17. And if in all respects unequal, be capable of eluding him, for a small force is but booty for one more powerful.

18. Now the general is the protector of the state. If this protection is all-embracing, the state will surely be strong; if the defective, the state will certainly be weak.

19. Now there are three ways in which a ruler can bring misfortune upon his army:

20. When ignorant that the army should not advance, to order an advance or ignorant that it should not retire, to order a retirement. This is described as "hobbling the army."

21. When ignorant of military affairs, to participate in their administration. This causes the officers to be perplexed.

22. When ignorant of command problems to share in the exercise of responsibilities. This engenders doubts in the minds of the officers.

23. If the army is confused and suspicious, neighboring rulers will cause trouble. This is what is meant by the saying: "A confused army leads to another's victory."

24. Now there are five circumstances in which victory may be predicted.

25. He who know when he can fight and when he cannot will be victorious.

26. He who understands how to use both large and small forces will be victorious.

27. He whose ranks are united in purpose will be victorious.

28. He who is product and lies in wait for an enemy who is not, will be victorious.

29. He whose generals are able and not interfered with by the sovereign will be victorious.

30. It is in these five matters that the way to victory is known.

31. Therefore I say: "Know the enemy and know yourself; in a hundred battles you will never be in peril."

32. When you are ignorant of the enemy but know yourself, your chances of winning or losing are equal.

33. If ignorant both of your enemy and of yourself, you are certain in every battle to be in peril.

Cyber Corollaries

1.   Generally in war, the best policy is to take a state intact; to ruin it is inferior to this.

Organizations must be prepared for the ebb and flow in cyber operations. The question is not which is the strongest, offense or defense. The question is what will safeguard the environment, protect the things of value, and sustain operations in the face of a cyber attack.

Paraphrasing *The Art of War*, victory needs to be achieved in the shortest possible time and with the least possible cost in lives and effort. For many nation-states and non-state actors, cyber conflict provides a strategic opportunity to change the playing field. For the goal in cyber conflict is not the physical destruction of the enemy; the goal is the disruption of communications, the elimination of the ability to respond in a coordinated fashion, the theft of goods, and/or the collapse of morale.

For most cyber operations, victory in cyber is not the defeat of a putative enemy. Rather, victory is safeguarding data and the successful navigation of and continual operations in a hostile cyber environment. Depending upon the cyber actors' motivations, this may or may not be true. For the "digital thieves," the low-level criminals or the highly organized syndicates, financially motivated cyber actors want to continue to put their hands into the virtual cookie jar. It is generally in their best, long-term interests to allow organizations to continue to operate as they siphon off their gains.

Organizations must be prepared for the ebb and flow in cyber operations. The question is not which is the strongest, offense or defense. The question is what will safeguard the environment, protect the things of value, and sustain operations in the face of a cyber attack.

> 4.   *Thus, what is of supreme importance in war is to attack the enemy's strategy.*

Another question for organizations is what will enable the achievement of the objectives and to gain the maximum advantage at the minimal cost. Cyber response starts with a well-organized plan. Sun Tzu advises a focus on strategies and tactics suitable to the cyber conditions, patience in waiting for suitable conditions, and the discipline to execute again and again.

Organizations need to have a cyber strategy to protect their own strategy and to thwart an adversary's strategy. The organization's strategy needs to consider all the cyber axioms.

By understanding what the attackers value—their motivation—organizations can disrupt them. If the motivation is financial, then the organization can ally with other entities to reduce the value of the target by intentionally compromising its integrity or prevent the distribution of the asset to reduce the attackers' market. If the motivation is political, then the organization can work to appropriately counter that.

> 5.   *Next best is to disrupt his alliances.*

If attacking a strategy did not work, Sun Tzu then advocated disrupting alliances by either preventing them from occurring or by causing them to fall apart. Defensive alliances include industry alliances, such as SWIFT; technology alliances, where vendors and organizations partner to create robust, secure systems; government alliances, where private entities and governments share threat information. These alliances can initially allow for information sharing and eventually grow to the adoption of common procedures and technology standards.

Active discouragement would be the intentional disruption of the attackers' alliances. In cyber, no actor is alone. The networks provide the pathway to the organization's door. The financial systems provide the mechanisms for payment. Organizations and governments need to work to systematically disable the attackers' own alliances.

IV: Disposition

The Stage

The Chinese character used in the title, *Hsing,* can be translated in many ways—*shape, form, appearance, formation,* or *disposition.* Since this chapter outlines how to use competitive positions, the translation "Disposition" is most appropriate.

Disposition is a military's organization, structure, strength, and placement of personnel, units, and assets in support of operation. In current American military jargon, *disposition* would be the order of battle.

Cyber operations are based on the current disposition, positions determined by previous projects, infrastructures, and investment. By recognizing its environment, organizations can plan and prepare to recognize opportunities, and influence their cyber environment, which should reduce their cyber weaknesses.

Almost all cyber environments and operations use existing infrastructure, the public marketplace, and previous investments. Appropriating the image of land where there is no need to remodel or demolish existing structures, technologists use the phrase *greenfield* to describe a project that does not have the constraints of prior work. Rarely does an organization have a greenfield. Organizations work with what they have.

SUN TZU said

1. Anciently the skillful warriors first made themselves invincible and awaited the enemy's moment of vulnerability.

2. Invincibility depends on one's self; the enemy's vulnerability on him.

3. It follows that those skilled in war can make themselves invincible but cannot cause an enemy to be certainly vulnerable.

4. Therefore it is said that one may know how to win, but cannot necessarily do so.

5. Invincibility lies in the defense; the possibility of victory in the attack.

6. One defends when his strength is inadequate; he attacks when it is abundant.

7. The experts in defense conceal themselves as under the ninefold earth; those skilled in attack move as from above the ninefold heavens. Thus, they are capable of both protecting themselves and gaining a complete victory.

8. To foresee a victory which the ordinary man can foresee is not the acme of skill;

9. To triumph in battle and be universally acclaimed "Expert" is not acme of skill, for to lift an autumn down requires no great strength; to distinguish between the sun and moon is not test of vision; to hear the thunderclap is no indication of acute hearing.

10. Anciently, those called skilled in war conquered an enemy easily conquered.

11. And therefore the victories won by a master of war gain him neither reputation for wisdom nor merit for valor.

12. For he wins his victories without erring. "Without erring" means that whatever he does ensures his victory; he conquers an enemy already defeated.

13. There the skillful commander takes up a position in which he cannot be defeated and misses no opportunity to master his enemy.

14. Thus a victorious army wins its victories before seeking battle; an army destined to defeat fights in the hope of winning.

15. Those skilled in war cultivate the Tao and preserve the laws and are therefore able to formulate victorious policies.

16. Now the elements of the art of war are first, measurement of space; second, estimation of quantities; third, calculations; fourth, comparison; and fifth, chances of victory.

17. Measurements of space are derived from the ground.

18. Quantities derive from measurement, figures from quantities, comparisons from figures, and victory from comparisons.

19. Thus a victorious army is as a hundredweight balanced against a grain; a defeated army as a grain balanced against a hundredweight.

20. It is because of disposition that a victorious general is able to make his people fight with the effect of pent-up waters which, suddenly released, plunge into a bottomless abyss.

Cyber Corollaries

> 2.   *Invincibility depends on one's self; the enemy's vulnerability on him.*

As seen, this chapter explains how use competitive positions. Individuals, businesses, organizations, and governments want to protect those things they hold dear, while cyber adversaries see that value as a target to exploit. Financial data, personal information, and intellectual property are just a few of the things held dear by their respective owners.

Following Sun Tzu's admonition to know yourself, as organizations gain understanding of their environments and of their cyber strengths and weaknesses, their ability to reduce their own vulnerabilities increases. Attackers operate in gray zones. Their threat can be reduced through recognition, remediation, and response. Since they are motivated by either financial or political considerations, organizations and their alliances can direct their responses to those motivations and disrupt them.

> 5.   *Invincibility lies in the defense; the possibility of victory in the attack.*

The traditional cyber defense is a defense-in-depth strategy. This strategy aligns security from the outside in. There are more safeguards at the border than there are in the interior.

Given the unrestricted and deceptive nature of cyber conflict, this traditional approach is insufficient on its own. A "no-trust" architecture embeds security throughout the organization and protects data at rest, in motion, and in transit on networks, in data centers, and across transmission systems.

The organization and the architecture separate duties of system administrators internally to ensure that no one individual holds all of the organization's cyber keys. They separate identified and authenticated traffic from anonymous traffic. The infrastructure is cloaked, obfuscated, and abstracted to prevent attackers from being able to perform any type of vulnerability scanning.

> 7.   *The experts in defense conceal themselves as under the ninefold earth; those skilled in attack move as from above the ninefold heavens. Thus, they are capable of both protecting themselves and gaining a complete victory.*

An enemy cannot kill what they cannot find. They cannot counter what they cannot understand. If the enemy cannot find and reach the target, they cannot act. Cyber concealment, obfuscation, deception, and the emplacement of obstacles are all part of the planning and baselining process and serve to help safeguard the environment.

> 16.   *Now the elements of the art of war are first, measurement of space; second, estimation of quantities; third, calculations; fourth, comparison; and fifth, chances of victory.*

> and

> 18.   *Quantities derive from measurement, figures from quantities, comparisons from figures, and victory from comparisons.*

Prescient of Walter Deming and big data analytics, Sun Tzu advised organizations to measure "everything" within their environment. It is through the gathering of metrics and their effective analysis that an organization prepares for conflict.

In the cyber environment, data is everywhere, and organizations are being propelled into the exabyte era. As Galileo said, "Measure what is measurable, and make measurable what is not so." It is about the importance of quantification as part of the scientific method. Organizations require awareness of and access to the data and the metrics to track the performance of their operations and financial management and for provisioning and safeguarding of services. The metrics strategy must fit the organization and conform to the available technology.

Cyber metrics are not a single form of data. Rather, the data collected and analyzed needs to represent an enterprise view and includes all of the various assets. As such, this strategy considers system-based data, web content, documents and records, images, messages, network logs, intrusion alerts, and other forms of relevant data.

Stewardship of the source data is needed, for the metrics that are closely connected to the core business, technology, and security functions can potentially be compromised by adversaries.

V: *Force*

The Stage

The title in the original Chinese is *Shih*, which can mean *energy*, *authority*, *influence*, and *force*. All these translations are appropriate.

Recall, the Warring States period was a time of innovation. Innovation occurred in economics, politics, warfare, and especially around philosophy. The philosophies of Confucianism, Taoism, and Legalism took form during this period. The Taoist concept of *chi* became familiar, even commonplace.

Chi is thought to be the life energy contained within matter. The notion of *chi* and its applications are embedded throughout Chinese culture.

As is well established elsewhere, there is an underlying current of Taoism, and in turn *chi*, throughout *The Art of War*. Chapter 5, "Force," explores and combines two distinct ideas into the application of *chi* in warfare: energy and momentum.

The Art of War provides insights on how to allow *chi's* energy and momentum assist a commander in achieving their objectives. Sun Tzu does not describe how to control *chi*. Rather, he provides indicators on how to recognize energy and channel momentum.

Information technology and, in turn, the cyber environment are disruptive and dynamic. New technologies are always emerging, and legacy technologies and processes are becoming obsolete at ever-faster rates. Virtually every organization, regardless of industry or size, has embraced information technology as a way of gaining a competitive edge. Cost optimization, competition, and a well-informed but demanding customer base have given organizations an impetus to invest in the new technologies.

These are cyber's energy and momentum. Organizations need to recognize the cyber trends impacting them as early as possible and channel the momentum to serve their objectives.

SUN TZU said

1. Generally, management of many is the same as management of few. It is a matter of organization.

2. And to control many is the same as to control few. This is a matter of formations and signals.

3. That the army is certain to sustain the enemy's attack without suffering defeat is due to operations of the extraordinary and the normal forces.

4. Troops thrown against the enemy as a grindstone against eggs is an example of a solid acting upon a void.

5. Generally, in battle, use the normal force to engage; use the extraordinary to win.

6. Now the resources of those skilled in the use of extraordinary forces are as infinite as the heavens and earth; as inexhaustible as the flow of the great rivers.

7. For they end and recommence; cyclical as are the movements of the sun and moon. They die away and are reborn; recurrent, as are the passing seasons.

8. The musical notes are only five in number but their melodies are so numerous that one cannot hear them all.

9. The primary colors are only five in number but their combinations are so infinite that one cannot visualize them all.

10. The flavors are only five in number but their blends are so various that one cannot taste them all.

11. In battle, there are only the normal and extraordinary forces, but their combinations are limitless; none can comprehend them all.

12. For these two forces are mutually reproductive; their interaction as endless as that of interlocked rings. Who can determine where one ends and the other begins?

13. When torrential water tosses boulders, it is because of its momentum;

14. When the strike of a hawk breaks the body of its prey, it is because of its timing.

15. Thus, the momentum of one skilled in war is overwhelming and his attack precisely regulated.

16. His potential is that of a fully drawn crossbow; his timing the release of the trigger.

17. In the tumult and uproar, the battle seems chaotic, but there is no disorder; the troops appear to be milling about in circles but cannot be defeated.

18. Apparent confusion is a product of good order; apparent cowardice, of courage; apparent weakness, of strength.

19. Order or disorder depends on organization; courage or cowardice on circumstances; strength or weakness on dispositions.

20. Thus, those skilled at making the enemy move do so by creating a situation to which he must conform; they entice him with something he is certain to take, and with lures of ostensible profit they await him in strength.

21. Therefore a skilled commander seeks victory from the situation and does not demand it of his subordinates.

22. He selects his men and they exploit the situation.

24. He who relies on the situation uses his men in fighting as one rolls logs or stones. Now the nature of logs and stones is that on stable ground they are static; on unstable ground, they move. If square, they stop; if round, they roll.

25. Thus the potential of troops skillfully commanded in battle may be compared to that of round boulders which roll down from mountain heights.

Cyber Corollaries

13.   When torrential water tosses boulders, it is because of its momentum;

and

14.   When the strike of a hawk breaks the body of its prey, it is because of its timing.

and

15.   Thus, the momentum of one skilled in war is overwhelming and his attack precisely regulated.

The text of this chapter provides indicators on how to recognize energy and channel momentum. Sun Tzu advises allowing natural force to assist commanders in achieving their objectives.

A slightly different way to think of force is as a trend. Trends are the general direction of a market, asset price, or technology. They can vary in length from short to intermediate to long, secular trends.

Sun Tzu clearly advises military leaders to look for the trends and to work with the trends. Caution should be taken if any position relies upon an existing trend reversing itself.

As the phrase goes, "A trend is your friend," or maybe better, "A recognized trend is your friend." Technology consumerization and mobility are examples of technology booms that have impacted cyber and created trends. Organizations need to learn how to identify trends and ride momentum.

VI: *Weaknesses & Strengthens*

The Stage

To explain the nature of success in conflict, "Weaknesses & Strengths" provides a process of understanding competitive weaknesses and strengths. Sun Tzu offers four axioms: use an enemy's weakness against him; use an enemy's apparent strengths against him; turn your own apparent weaknesses into strength; and use stratagem to mask weakness and exploit strengths.

Many business organizations have used a "Strengths—Weaknesses—Opportunities—Threats" (SWOT) analysis to help guide their efforts. A traditional business SWOT analysis examines the internal and external attributes and resources supporting or opposing a desired outcome.

Weaknesses and strengthens are the two sides of the cyber coin. The question becomes how to recognize weaknesses and strengths and how to respond accordingly. In cyber conflict, every individual, organization, corporation, and government has strengths and weakness, each one of them.

To fully understand their cyber environment, organizations need to conduct a cyber SWOT analysis using the cognitive axioms previously noted in concert with the recognize-remediate-respond approach as detailed in chapter 7.

Adversaries focus on identifying weaknesses, believing that success can only be achieved by taking advantage of those weaknesses. Organizations need to counter those conditions to ensure that they can successfully maneuver through the cyber environment. Overly complex technology, supply chain dependencies, insufficient funding, and vulnerable communications networks are just some of the readily exploitable cyber weaknesses that can be identified and proactively resolved.

SUN TZU said

1. Generally, he who occupies the field of battle first and awaits his enemy is at ease; he who comes later to the scene and rushes into the fight is weary.

2. And therefore those skilled in war bring the enemy to the field of battle and are not brought there by him.

3. One able to make the enemy come of his own accord does so by offering him some advantage. And one able to prevent him from coming does so by hurting him.

4. When the enemy is at ease, be able to weary him; when well fed, to starve him; when at rest, to make him move.

5. Appear at places to which he must hasten; move swiftly where he does not expect you.

6. That you may march a thousand *li* without wearying yourself is because you travel where there is no enemy.

7. To be certain to take what you attack is to attack a place the enemy does not protect. To be certain to hold what you defend is to defend a place the enemy does not attack.

8. Therefore, against those skilled in attack, an enemy does not know where to defend; against the experts in defense, the enemy does not where to attack.

9. Subtle and insubstantial, the expert leaves no trace; divinely mysterious, he is inaudible. Thus he is master of his enemy's fate.

10. He whose advance is irresistible plunges into his enemy's weak positions; he who in withdrawal cannot be pursued moves so swiftly that he cannot be overtaken.

11. When I wish to give battle, my enemy, even though protected by high walls and deep moats, cannot help but engage me, for I attack a position he must succor.

12. When I wish to avoid battle I may defend myself simply by drawing a line on the ground; the enemy will be unable to attack me because I divert him from going where he wishes.

13. If I am able to determine the enemy's dispositions while at the same time I conceal my own then I can concentrate and he must divide. And if I concentrate while he divides, I can use my entire strength to attack a fraction of his. There, I will be numerically superior. Then, if I am able to use many to strike few at the selected point, those I deal with will be in dire straits.

14. The enemy must not know where I intend to give battle. For if he does not know where I intend to give battle he must prepare in a great many places. And when he prepares in a great many places, those I have to fight in any one place will be few.

15. For if he prepares to the front his rear will be weak, and if to the rear, his front will be fragile. If he prepares to the left, his right will be vulnerable and if to the right, there will be few on his left. And when he prepares everywhere he will be weak everywhere.

16. One who has few must prepare against the enemy; one who has many makes the enemy prepare against him.

17. If one knows where and when a battle will be fought his troops can march a thousand *li* and meet on the field. But if one knows neither the battleground nor the day of the battle, the left will be unable to aid the right, or the right, the left, the van to support the rear, or the rear, the van. How much more is this so when separated by a several tens or *li*, or indeed, by even a few!

18. Although I estimated the troops of Yüeh as many, of what benefit is this superiority in respect to the outcome?

19. Thus I say that victory can be created. For even if the enemy is numerous, I can prevent him from engaging.

20. Therefore, determine the enemy's plans and you will know which strategy will be successful and which will not;

21. Agitate him and ascertain the pattern of his movement.

22. Determine his dispositions and so ascertain the field of battle.

23. Probe him and learn where his strength is abundant and where deficient.

24. The ultimate in disposing one's troops is to be without ascertainable shape. Then the most pene-

trating spies cannot pry in nor can the wise lay plans against you.

25. It is according to the shape that I lay the plans for victory, but the multitude does not comprehend this. Although everyone can see the outward aspects, none understands the way in which I have created victory.

26. Therefore, when I have won a victory, I do not repeat my tactics, but respond to circumstances in an infinite variety of ways.

27. Now an army be likened to water, for just as flowing water avoids the heights and hastens to the lowlands, so an army avoids strengths and strikes weaknesses.

28. And as water shapes its flow in accordance with the ground, so an army manages its victory in accordance with the situation of the enemy.

29. And as water has no constant form, there are in war no constant conditions.

30. Thus, one able to gain the victory by modifying his tactics in accordance with the enemy situation may be said to be divine.

31. Of the five elements, none is always predominate; of the four seasons, none lasts forever; of the days, some are long and some short, and the moon waxes and wanes.

Cyber Corollaries

> 2.   *And therefore those skilled in war bring the enemy to the field of battle and are not brought there by him.*

The way to achieve success in cyber conflict is to be flexible and to adapt strategies and tactics to the prevailing and expected conditions within the context of the plan. While direct conflict is best avoided, this cannot always be done. In those situations, organizations must understand how to tip the environmental scale in their favor.

Using technology obstacles and cover, organizations can direct attackers in a specific direction. With decoy networks, systems, or servers, the so-called honeypot networks, organizations can either misdirect or gather information regarding their adversaries. The no-trust architecture emplaces obstacles throughout the environment, diverting attackers in specific directions. The routine destruction of the internal technology environment eliminates potential hiding locations.

> 3.   *One able to make the enemy come of his own accord does so by offering him some advantage. And one able to prevent him from coming does so by hurting him.*

This stanza advises not to attack a well-arrayed enemy and to avoid the opponent's strengths. An understanding of their cyber environment—through the right approach of recognizing, remediating, and responding—lets organizations know their vulnerabilities and apply additional resources. Further, recognition allows organizations to being the response process outlined in chapter 7 and to actively discourage malicious actors. Active discouragement can reduce the potential value of the targets or creating obstacles that dissuade malicious actions. If the potential target is tainted, then it does not have the same value.

Sun Tzu advised organizations to threaten something that the enemy must protect, a weak point or something of value. Moral considerations aside, this advice requires taking advantage of an opponent's exploitable weakness.

> 14.   The enemy must not know where I intend to give battle. For if he does not know where I intend to give battle, he must prepare in a great many places. And when he prepares in a great many places, those I have to fight in any one place will be few.

Organizations should use this adage in the reverse. Through preparation, tactics, disposition, and knowledge of weaknesses and strengths, they can cause cyber attackers to either pause or turn away by forcing them to engage in many places and to engage in ways not of the attacker's choosing. If the attacker does not know where an organization intends to defend or how it intends to defend, the aggressor will be forced to attack without a meaningful way to control the outcome.

Organizations can routinely restart their environments, essentially destroying their existing infrastructures on a periodic basis and starting every environment from scratch. It is possible to create, modify, and destroy an organization's infrastructure quickly and frequently. Security controls should be dynamic, which causes an attacker to probe and assault in many places, possibly in manners that expose them. The use of the newer network management techniques allows for the network design to be modified as rapidly. In short, an organization's targets and key terrain can be instantaneously moved in the face of a network attack.

> 23.   Probe him and learn where his strength is abundant and where deficient.

Weaknesses and strengths are continually being tested. In the unrestricted world of cyber conflict, the rules are never completely set; new weapons and technology are deployed; the conflict across all facets of the organization expanding as new technologies become weaknesses to exploit.

To counter adversarial probes, organizations need to conduct meaningful independent and objective vulnerability assessments and penetration testing. They need to see what the malicious actors see and, in turn, recognize what needs to be protected. Then they can apply the resources necessary to safeguard the environment. A detailed view of the threats enables the organization or the corporation to safely protect its data.

VII: Maneuvers

The Stage

This chapter, "Maneuvers," also commonly translated as "Armed Conflict," discusses the art of maneuvering an army. It also cautions about the dangers of direct conflict.

While direct conflict is best avoided, this cannot always be done. In those situations, a commander must understand how to tip the balance in his or her favor.

As previously noted, tactics are vital in conducting operations. Tactics include the process of planning, coordinating, and giving general directions to all operations with the aim of meeting military, political, and/or economic objectives. Maneuver involves the use of those tactics to deploy the resources.

Today's cyber conflict is significantly more complex than Sun Tzu's land warfare due to a variety of factors—the digital pace, the omnidirectional nature, the atemporal nature, the asymmetric deployment of forces. Cyber is a conflict without boundaries and where threats appear from every direction and at any time. It is a conflict where the relative military power differs significantly among the various actors. It is a conflict where machines battle machines with millisecond response time.

Using the analysis put forward by Scott D. Applegate in his paper "The Principle of Maneuver in Cyber Operations," offensive cyber maneuvers have several characteristics:

- Speed: Operations are conducted in Internet speed, with rapid, almost instantaneous actions across broad areas.

- Operational reach: With its highly aspatial nature, cyber actors can reach across the globe with a simple click, while concurrently conducting local operations.

The physics of location is all but eliminated. Actors have become nearly "omnipresent."

- Access and control: Actors require access to and control of systems to effectively operate. These systems could be friendly, neutral, or enemy.

- Deception and limited attribution: As noted in the cognitive principles, cyber conflict inherently involves deception. Without the traditional cues, deception has become rampant. Impersonation and system hijacking abound. Further, given the open nature of the technology architecture and the original premise of implicit trust, the ability to know the enemy is missing.

- Rapid concentration: Unlike traditional warfare where it takes time to concentrate physical troops or move aircraft or ships into position, the virtual warriors found within the cyber environment can form and reform within seconds.

- Non-serial and distributed: There is an atemporal nature to the cyber environment and cyber attacks, attackers, and defenders.

These are the atemporal, unrestricted maneuvers that organizations must prepare for. Cyber defensive maneuvers follow the maneuvers like those use in traditional warfare: perimeter defenses, intrusion detection, and defense in depth. However, implementation and the need to continually adapt make defensive cyber maneuvers a great deal more difficult than the maneuvers laid out in *The Art of War*.

SUN TZU said

1. Normally, when the army is employed, the general first receives his commands from the sovereign. He assembles the troops and mobilizes the people. He blends the army into a harmonious entity and encamps its.

2. Nothing is more difficult than the art of maneuver. What is difficult about maneuver is to make the devious route the more direct and to turn misfortune into advantage.

3. Thus, march by an indirect route and divert the enemy by enticing him with a bait. So doing, you may set out after he does so and arrive before him. One able to do this understands the strategy of the direct and the indirect.

4. Now both advantage and danger are inherent in maneuver.

5. One who sets the entire army in motion to chase an advantage will not attain it.

6. If he abandons the camp to contend for advantage the stores will be lost.

7. It follows that when one rolls up the armor and sets out speedily, stopping neither day nor night and marching at double time for a hundred *li*, the three commanders will be captured. The vigorous troops will arrive first and the feeble straggle along behind, so that if this method is used only one-tenth of the army will arrive.

8. In a forced march of fifty *li*, the commander of the van will fall, and using this method but half the army will arrive. In a forced march of thirty *li*, but two-thirds will arrive.

9. It follows that an army which lacks heavy equipment, fodder, food and stores will be lost.

10. Those who do not know the conditions of mountains and forest, hazardous defiles, marshes and swamps, cannot conduct the march of an army;

11. Those who do not use local guides are unable to obtain the advantages of the ground.

12. Now war is based on deception. Move when it is advantageous and create changes in the situation by dispersal and concentration of forces.

13. When campaigning, be swift as the wind; in leisurely march, majestic as the forest; in raiding and plundering, like fire; in standing, firm as the mountain. As unfathomable as the clouds, move like a thunderbolt.

14. When you plunder the countryside, divide your forces. When you conquer territory, divide the profits.

15. Weigh the situation, then move.

16. He who knows the art of the direct and the indirect approach will be victorious. Such is the art of maneuvering.

17. *The Book of Military Administration* says: "As the voice cannot be heard in battle, drums and bells are used. As troops cannot see each other clearly in battle, flags and banners are used."

18. Now gongs and drums, banner and flags are used to focus the attention of the troops. When the troops can be thus united, the brave cannot advance alone, nor can the cowardly withdraw. This is the art of employing a host.

19. In night fighting use many torches and drums, in day fighting many banners and flags in order to influence the sight and hearing of our troops.

20. Now an army may be robbed of its spirt and its commander deprived of his courage.

21. During the early morning spirts are keen, during the day they flag, and in the evening thoughts turn toward home.

22. And therefore those skilled in war avoid the enemy when his spirit is keen and attack him when it is sluggish and his soldiers homesick. This is control of the moral factor.

23. In good order they await a disorderly enemy; in serenity, a clamorous one. This is control of the mental factor.

24. Close to the field of battle, they await an enemy coming from afar; at rest, an exhausted enemy; with well-fed troops, hungry ones. This is control of the physical factor.

25. They do not engage an enemy advancing with well-ordered banners nor one whose formations are in impressive array. This is control of the factor of changing circumstances.

26. There, the art of employing troops is that when the enemy occupies high ground, do not confront him; with his back resting on hills, do not oppose him.

27. When he pretends to flee, do not pursue.

28, Do not attack his elite troops.

29. Do not gobble proffered baits.

30. Do not thwart an enemy returning homewards.

31. To surround an enemy you must leave a way of escape.

32. Do not press an enemy at bay.

33. This is the method of employing troops.

Cyber Corollaries

> *2.   Nothing is more difficult than the art of maneuver. What is difficult about maneuver is to make the devious route the more direct and to turn misfortune into advantage.*

For Sun Tzu, the objective of warfare is not the enemy's total destruction. The objective of warfare is "winning all without fighting." The way to achieve success in cyber conflict is to be flexible in adapting strategies and tactics to conditions within the context of the environment. Organizations must understand how to tip the balance in their favor by "being faster than the slowest runner" or understanding how to mitigate enough risk so that the attacker will move on to an easier target.

Only through understanding the nature of cyber conflict and its unrestricted use of weakness and deception to target vulnerable spots can an organization understand how to keep its plans and strategy closely held and to keep its terrain and potential targets secured. Through adaptable defensive tactics and maneuvers that deceive the adversary about one's true intentions, organizations can maneuver through the cyber environment.

> *12.   Now war is based on deception. Move when it is advantageous and create changes in the situation by dispersal and concentration of forces.*

Common in the physical world, deception is rampant in the cyber environment. As society has become more connected and complex, it has become more vulnerable. Whether actually connected to the network or simply near it, nearly every so-called smart device expands the threat surface well beyond anyone's ability to control. These devices can readily impersonate authorized uses, transmitting e-mails to gain passwords, or sending links to false website where a user is asked to enter personal data. Through network obfuscation, data encryption, and dynamic security controls, organizations can change their situation and cause the attackers to move on.

> 17.   The Book of Military Administration *says: "As the voice cannot be heard in battle, drums and bells are used. As troops cannot see each other clearly in battle, flags and banners are used.*

and

> 18.   *Now gongs and drums, banner and flags are used to focus the attention of the troops. When the troops can be thus united, the brave cannot advance alone, nor can the cowardly withdraw. This is the art of employing a host.*

With an ever-increasing number of sensors and input devices, with the proliferation of agents, logs, and drones feeding data into analytic engines, the amount of noise within the cyber environment has turned into a digital tidal wave, a tidal wave that floods the analysts and organizations with meaningless information, drowning out their ability to focus in on resolving the actual issues. Through deception and disinformation, adversaries will add noise into the environment to prevent organizations from clearly seeing the environment.

The numerous data sources are like individual voices. When conversations increase by multiples of volume, variety, and velocity, there is almost no way to pick out one voice. It all turns into white noise.

The volume of the information in the cyber environment is almost overwhelming. Every click, data stream, sensor log, call, and video adds to the variety of information. The speed of input, its velocity, has dramatically increased. All three vectors have increased exponentially.

There is a basic confusion and inability to recognize what matters and how to focus an organization on what matters.

With an organizational strategy and plan to execute on recognize-remediate-respond, organizations can filter the noise and develop an integrated approach to respond to the cyber challenges.

> 21. *During the early morning spirts are keen, during the day they flag, and in the evening thoughts turn toward home.*

In this stanza, Sun Tzu clearly recognized the natural rhythms of the day, an early understanding that humans have a circadian rhythm. In the current cyber conflict, there is an implicit expectation that the team is constantly awake, vigilant for alerts and attacks. Through its state of endless watchfulness and always being on, cyber conflict upends the natural rhythm. This vigilance causes a circadian rhythm misalignment, with an increase in the potential for mistakes. It has been shown that these misalignments affect decision making as well as psychomotor skills. In the zero-defect environment of cyber conflict, those mistakes can be very dangerous.

VIII: *Adapting*

The Stage

Setting the stage for the next three, slightly longer chapters, this chapter, traditionally entitled "The Nine Variables," focuses on the need to adapt to the conditions encountered. While the following three chapters provide more proscriptive approaches, this chapter puts forward the broader notion that every circumstance is distinct, but that every circumstance is based upon familiar, common elements. Responses can be creative and flexible; however, they still need to adhere to the rules of "standard response." Further, responses cannot be taken out of ignorance and without accounting for the second-, third-, and fourth-order effects.

In almost every era, military tactics and business operations have changed. Change is due to political, economic, cultural, and technological advances.

As was seen in the Warring States period, innovation occurred and the culture adapted. A revolution in weapons development preceded the evolution of military theory.

The same holds true for cyber conflict. The cyber environment—tactics, motivations, technology, actions—are constantly changing. Organizations need to adapt so that they can complete their mission in the face of cyber conflict.

Organizations need to be able to operate through and recover from those attacks. Success in cyber conflict is not about eliminating all the threats but understanding the environment, resolving the known threats, and adapting to each new threat.

Adaptability—resiliency—is about using a risk-management approach that includes people, processes, and technology. The correlation of security logs and threat intelligence is important. What is more important is having the flexibility in the people, technologies, laws, regulations, and policies to rapidly recognize-remediate-respond. The complex omnidirectional, unrestricted cyber attacks can compromise entire technology infrastructures in a matter of minutes. Critical functions need to be maintained as these breaches occur. *The Art of War* describes responses that need to be creative and flexible. Concurrently, to allow organizations to move rapidly, the responses need to be standardized, repeatable processes, a.k.a. *industrialized remediation*.

SUN TZU said

1. In general, the system of employing troops is that the commander receives his mandate from the sovereign to mobilize the people and assemble the army.

2. You should not encamp in low-lying ground.

3. In communicating ground, unite with your allies.

4. You should not linger in desolate ground.

5. In enclosed ground, resourcefulness is required.

6. In death ground, fight.

7. There are some roads not to follow; some troops not to strike; some cities not assault; and some ground which should not be contest.

8. There are occasions when the commands of the sovereign need not be obeyed.

9. A general thoroughly versed in the advantages of the nine variable factors knows how to employ troops.

10. The general who does not understand the advantages of the nine variable factors will not be able to use the ground to his advantage even though familiar with it.

11. In the direction of military operations one who does not understand the tactics suitable to the nine variable situations will be unable to use his

troops effectively, even if he understands the "five advantages."

12. And of this reason, the wise general in his deliberations must consider both favorable and unfavorable factors.

13. By taking into account the favorable factors, he makes his plan feasible; by taking into account the unfavorable, he may resolve the difficulties.

14. He who intimidates his neighbors does so by inflicting injury upon them.

15. He wearies them by keeping them constantly occupied, and makes them rush about by offering them ostensible advantages.

16. It is a doctrine of war not to assume the enemy will not come, but rather to rely on one's readiness to meet him; not to presume that he will not attack, but rather to make one's self invincible.

17. There are five qualities which are dangerous in the character of a general.

18. If reckless, he can be killed;

19. If cowardly, captured;

20. If quick-tempered you can make a fool of him;

21. If he has too delicate a sense of honor, you can make false and defamatory statements about him;

22. If his is of a compassionate nature you can harass him.

23. Now these five traits of character are serious faults in a general and in military operations are calamitous.

24. The ruin of an army and the death of a general are inevitable results of these shortcomings. They must be deeply pondered.

Cyber Corollaries

4.   *You should not linger in desolate ground.*

As demonstrated by incidents of cyber espionage and data leaks, firewalls and static defensives are vulnerable. Operational resiliency, industrialized cyber solutions, and effective training are part of an organization's approach to ensuring true security rather than merely complying with security checklists.

> 16.   *It is a doctrine of war not to assume the enemy will not come, but rather to rely on one's readiness to meet him; not to presume that he will not attack, but rather to make one's self invincible.*

Because they lack the flexibility to respond to unforeseen difficulties, conventional static security approaches are routinely ineffective. There is no such thing as a perfect strategy or plan. Devising strategies involves striking a series of compromises. Effective defense cyber deposition requires a fundamental rethinking of the traditional approaches to ensure that defenses become more flexible in the face of cyber conflict. Effective cyber defense also requires the acknowledgment of the nontraditional nature of the adversaries' command system and the difficulty in responding to its rhizomatic nature.

IX: Situational Awareness

The Stage

The title of this chapter is traditionally translated as "Armed March" or "March," which does not accurately describe its contents. Current military theory uses the terms *situational awareness* and *situational assessment* to describe what Sun Tzu characterized as "armed march." This chapter describes the way to analyze the different conditions that an army finds as it moves into new competitive areas. It explains what various situations mean and how to respond to them.

Situational awareness is a state of knowledge. It is not only the perception of the environment and the events; it is also the understanding of the meaning of those elements.

In a similar fashion, *situational assessment* describes the processes used to achieve that knowledge by analysis procedures and mental models. Those models are routinely "a set of well-defined, highly organized, yet dynamic knowledge structures developed over time from experience." Accurate analysis procedures and comprehensive mental models are prerequisites for developing the situational awareness necessary to achieve success in cyber conflict.

Comprehensive cyber situational awareness directly uses the attributes of military and law enforcement theories. It begins with the four traditional military themes: identification, location, execution, and results; the law enforcement principles of safeguarding and community policing are also included. It then layers on cyber cognitive principles as it filters through an understanding of the cyber environment for the network and system components, the organization's mission, and the threat information.

Cyber situational awareness and its underlying processes use these attributes to make an educated and informed assessment of the cyber environment and the emerging threats and trends to determine what action to take in order to safeguard the organization.

SUN TZU said

1. Generally when taking up a position and confronting the enemy, having crossed the mountains, stay close to valleys. Encamp on high ground facing the sunny side.

2. Fight downhill; do not ascend to attack.

3. So much for taking position in mountains.

4. After crossing a river you must move some distance away from it.

5. When an advancing enemy crosses water do not meet him at the water's edge. It is advantageous to allow half his force to cross and then strike.

6. If you wish to give battle, do not confront you enemy close to the water. Take position on high ground facing the sunlight. Do not take position downstream.

7. This relates to taking up positions near a river.

8. Cross salt marshes speedily. Do not linger in them. If you encounter the enemy in the middle of a salt marsh you must take position close to grass and water with trees to your rear.

9. This has to do with taking up position in salt marshes.

10. In level ground occupy a position which facilitates your action. With heights to your rear and right, the field of battle is to the front and the rear is safe.

11. This is how to take up position in level ground.

12. Generally, these are advantageous for encamping in the four situations named. By using them the Yellow Emperor conquered four sovereigns.

13. An army prefers high ground to low; esteems sunlight and dislikes shade. Thus, while nourishing its health, the army occupies a firm position. An army that does not suffer from countless diseases is said to be certain of victory.

14. When near mounds, foothills, dikes, or embankments, you must take position on the sunny side and rest your right and rear of them.

15. These methods are advantageous for the army and gain the help the ground affords.

16. Where there are precipitous torrents, "Heavenly Wells," "Heavenly Prisons," "Heavenly Nets," "Heavenly Traps," and "Heavenly Cracks," you must march speedily away from them. Do not approach them.

17. I keep a distance from these and draw the enemy toward them. I face them and cause him to put his back to them.

18. When on the flanks of the army there are dangerous defiles or ponds covered with aquatic grasses where reeds and rushes grow, or forested mountains with dense tangled undergrowth you must carefully search them out, for these are places where ambushes are laid and spies are hidden.

19. When the enemy is nearby but lying low he is depending on a favorable position. When he challenges to battle from afar he wishes to lure you to advance, for when he is in easy ground he is in an advantageous position.

20. When the trees are seen to move the enemy is advancing.

21. When many obstacles have been placed in the undergrowth, it is for the purpose of deception.

22. Birds rising in flight is a sign that the enemy is lying in ambush; when the wild animals are startled and flee he is trying to take you unaware.

23. Dust spurting upward in high straight columns indicates the approach of chariots. When it hangs low and is widespread infantry is approaching.

24. When dust rises in scattered area the enemy is bringing in firewood; when there are numerous small patches which seem to come and go he is encamping the army.

25. When the enemy's envoys speak in humble terms, but he continues his preparations, he will advance.

26. When their language is deceptive but the enemy pretentiously advances, he will retreat.

27. When the envoys speak in apologetic terms, he wishes a respite.

28. When without a previous understanding the enemy asks for a truce, he is plotting.

29. When light chariots first go out and take position on the flanks the enemy is forming for battle.

30. When his troops march speedily and he parades his battle chariots he is expecting to rendezvous with reinforcements.

31. When half his force advances and half withdraws he is attempt to decoy you.

32. When his troops lean on their weapons, they are famished.

33. When drawers of water drink before carrying it to camp, his troops are suffering from thirst.

34. When the enemy sees an advantage but does not advance to seize it, he is fatigued.

35. When birds gather above his camp sites, they are empty.

36. When at night the enemy's camp is clamorous, he is fearful.

37. When his troops are disorderly, the general has no prestige.

38. When his flags and banners move about constantly he is in disarray.

39. If the officers are short-tempered they are exhausted.

40. When the enemy feeds grain to the horses and his men meat and when his troops neither hang up their cooking pots nor return to their shelters, the enemy is desperate.

41. When the troops continually gather in small groups and whisper together the general has lost the confidence of the army.

42. Too frequent rewards indicate that the general is at the end of his resources; too frequent punishments that he is in acute distress.

43. If the officers at first treat the men violently and later are fearful of them, the limit of indiscipline has been reached.

44. When the enemy troops are in high spirits and, although facing you, do not join battle for a long time, nor leave, you must thoroughly investigate the situation.

45. In war, numbers alone confer no advantage. Do no advance relying on sheer military power.

46. It is sufficient to estimate the enemy situation correctly and to concentrate your strength to capture him. There is no more to it than this. He who lacks foresight and underestimates his enemy will surely be captured by him.

47. If troops are punished before their loyalty is secured they will be disobedient. If not obedient, it is difficult to employ them. If troops are loyal but punishments are not enforced you cannot employ them.

48. Thus, command them with civility and imbue them uniformly with martial ardor and it may be said that victory is certain.

49. If orders which are consistently effective are used in instructing the troops, they will be obedient. If orders which are not consistently effective are used in instructing them, they will be disobedient.

50. When orders are consistently trustworthy and observed, the relationship of a commander with his troops is satisfactory.

Cyber Corollaries

13. An army prefers high ground to low; esteems sunlight and dislikes shade. Thus, while nourishing its health, the army occupies a firm position. An army that does not suffer from countless diseases is said to be certain of victory.

This particular stanza encapsulates a great deal—high ground, sunlight, and freedom from diseases. In the early days of combat, the high ground was a preferred location due to the tactical advantage it typically gave combat units. From the high ground, the unit could gain more range from its artillery and small arms. The unit would be able see farther. It is harder to attack uphill than it is to attack downhill. Signaling equipment was more effective. It was a better, safer place.

In cyber, there is no high ground in the traditional military sense. The omnidirectional, unrestricted nature of cyber conflict means that attacks can come from anywhere. With advanced persistent threats, the attacks could come from anytime—the past, present, and future. There is no safe zone.

In cybersecurity, sunshine means transparency. Transparency increases accountability and allows organizations to focus the tactics and maneuvers necessary to success in the cyber conflict. Through alliances, organizations can look at their community for examples of what works and what does not work and share cyber threat information. This transparency strengthens the entire community.

Adversaries are not just watching networks and end points to determine how they will attack. Adversaries are actively probing the environment, exploring digital shadows, identifying vulnerabilities, and launching attacks. Transparency will shine light on their activities before they gain a deep situational awareness of their opponents' cyber posture.

Viruses, malware, advanced persistent threats, denial of service attacks, these are some of many diseases found within the cyber environment. It is through a rigorous practice of cyber hygiene, the establishment, and maintenance of daily routines and practices that organizations can avoid cyber diseases. Some of the hygiene method include updating virus definitions and software, routine security scans, and encrypting personal data.

> 46.   It is sufficient to estimate the enemy situation correctly and to concentrate your strength to capture him. There is no more to it than this. He who lacks foresight and underestimates his enemy will surely be captured by him.

The current situational awareness models no longer suffice. Cyber conflict presents a problem for the development of situational awareness capacity. There is an enormous volume of data. This volume can overpower the capability of an experienced leader to process and will frequently paralyze a novice decision maker.

X: Targets & Terrain

The Stage

In this chapter, Sun Tzu analyzes natural and man-made features. In the central China region of the Warring States period, land warfare was the only form of warfare. This chapter is filled with detailed discussion about terrain, to include specific advantages and disadvantages. With rivers, plains, mountains, and gorges, geographical positions could offer protection from the other kingdoms. Water was an offensive or defensive obstacle. The other domains—air, space, and cyber—were not factors.

Cyber conflict poses different challenges for target and terrain identification. With its concurrent physical and virtual nature, the recognition of cyber target and terrain identification is difficult. Two questions need to be considered:

- What are cyber targets?

- What is key cyber terrain?

Cyber targets are the assets that an organization holds dear, the digital things of value. In short, targets are the organization's data. These can include personal information, personal correspondence, financial systems, medical information, intellectual property (IP), or command and control databases.

As defined by the U.S. military, *key cyber terrain* is the "physical and logical elements of the domain that enable mission essential functions." Key cyber terrain are the devices holding and pathways to the cyber targets. As other literature has identified, cyber terrain features can include network components and connections, systems components, perimeter defense devices, connections and interfaces, and spectrum. Of note, spectrum is a cyber terrain feature in the sense that those who can get their signal through and keep others from getting their signal will have a distinct advantage.

SUN TZU said

1. Ground may be classified according to its nature as accessible, entrapping, indecisive, constricted, precipitous, and distant.

2. Ground which both we and the enemy can traverse with equal ease is called accessible. In such ground, he who first takes high sunny positions convenient to his supply routes can fight advantageously.

3. Ground easy to get out of but difficult to return to is entrapping. The nature of this ground is such that if the enemy is unprepared and you sally out you may defeat him. If the enemy is prepared and you go out and engage, but do not win, it is difficult to return. This is unprofitable.

4. Ground equally disadvantageous for both the enemy and ourselves to enter is indecisive. The nature of this ground is such that although the enemy holds out a bait I do not go forth but entice him by marching off. When I have drawn out half his force, I can strike him advantageously.

5. If I first occupy constricted ground I must block the passes and await the enemy. If the enemy first occupies such ground and blocks the defiles I should not follow him; if he does not block them completely I may do so.

6. In precipitous ground I must take position on the sunny heights and await the enemy. If he first occupies such ground I lure him by marching off; I do not follow him.

7. When at a distance from an enemy of equal strength it is difficult to provoke battle and unprofitable to engage him in his chosen position.

8. These are the principles relating to six different types of ground. It is the highest responsibility of the general to inquire into them with the utmost care.

9. Now when troops flee, are insubordinate, distressed, collapse in disorder or are routed, it is the fault of the general. None of these disasters can be attributed to natural causes.

10. Other conditions being equal, if a force attacks one ten times its size, the result is flight.

11. When troops are strong and officers weak, the army is insubordinate.

12. When the officers are valiant and the troops ineffective the army is in distress.

13. When senior officers are angry and insubordinate, and on encountering the enemy rush into battle with no understanding of the feasibility of engaging and without orders from the commander, the army is in a state of collapse.

14. When the general is morally weak and his discipline not strict, when his instructions and guidance are not enlightened, when there are no consistent rules to guide the officers and men and when the formation are slovenly the army is in disorder.

15. When a commander unable to estimate his enemy uses a small force to engage a large one, or weak troops to strike the strong, or when he fails to select shock troops for the van, the result is rout.

16. When any of these six conditions prevails the army is on the road to defeat. It is the responsibility of the general that he examine them carefully.

17. Conformation of the ground is the greatest assurance in battle. Therefore, to estimate the enemy situation and to calculate distance and the degree of difficulty of the terrain so as to control victory are virtues of the superior general. He who fights with full knowledge of these factors is certain to win; he who does not will surely be defeated.

18. If the situation is one of victory but the sovereign has issued orders not to engage, the general may decide to fight. If the situation is such that he cannot win, but the sovereign has issued orders to engage, he need not do so.

19. And therefore the general who in advancing does not seek personal fame, and in withdrawing is not concerned with avoiding punishment, but whose only purpose is to protect the people and promote the best interests of his sovereign, is the precious jewel of the state.

20. Because such a general regards his men as infants they will march with him into the deepest valleys. He treats them as his own beloved sons and they will die with him.

22. If I know that my troops are capable of striking the enemy, but do not know that he is invulnerable to attack, my chance of victory is but half.

23. If I know that the enemy is vulnerable to attack, but do not know if my troops are incapable of striking him, my chance of victory is but half.

24. If I know that the enemy can be attacked and that my troops are capable of attacking him, but do not realize that because of the conformation of the ground I should not attack, my chance of victory is but half.

25. Therefore when those experienced in war move they make no mistakes; when they act, their resources limitless.

26. And therefore I say: "Know your enemy, know yourself; your victory will never be endangered. Know the ground, know the weather; your victory will then be total."

Cyber Corollaries

1.   Ground may be classified according to its nature as accessible, entrapping, indecisive, constricted, precipitous, and distant.

A response to cyber conflict requires both a static map and a dynamic overlay to fully understand the targets and terrain. The static map is the baseline that depicts both the cyber terrain features and the key targets. It lets an organization understand its cyber terrain in a manner like a river traveler exploring and mapping uncharted wilderness inhabited by hostile people. Like a river map, the analysis of network maps, system configurations, utilization trends, incident reports, and other profiling data shows where an organization has been and the obstacles that it has encountered in getting there.

The baseline records these experiences to gain a sense of the character and profile of the cyber river and its response patterns. This record enhances an organization's understanding of the best way to use the river, its cyber vulnerabilities and the hostile intent of the other actors.

However, given the inevitable twists and turns cyber takes, organizations cannot always see where the trends are going to carry them or how the other actors will impact them and require the dynamic overlay that situational awareness provides. An understanding of the terrain and the cyber targets shapes an organization's strategies and tactics, allowing them to prepare their plans and defenses.

XI: Terrain Analysis

The Stage

Terrain analysis, to understand the environment, is vital to success, and Sun Tzu, appropriately, spends more time on the topic than on any other. This is the last and the longest of the detailed chapters in *The Art of War*. In this chapter, Sun Tzu describes the common varieties of ground and provides analysis on how to conduct operations on that ground. He addresses the need to fully understand the impact of terrain on operations, especially when a force must attack an opponent that has had time to prepare a well-organized defense.

Current United States military doctrine has developed a structured approach to analyze terrain. It uses the principle of OCOKA, which stands for *observation, cover and concealment, obstacles, key terrain*, and *avenues of approach*.

Observation refers to the ability to see friendly and enemy forces from a vantage point. *Cover* is protection from the impact of hostile activities, while *concealment* is protection from observation. *Obstacles* are natural or man-made impediments. An *avenue of approach* is a ground or air route leading to an objective or the key terrain leading to an objective.

Similar to Sun Tzu's approach and United States military terrain analysis process, a cyber terrain analysis is conducted to identify key cyber terrain features, specifically for observation, cover and concealment, obstacles, and avenues of approach. A cyber terrain analysis discovers and classifies devices, correlates the devices, models vulnerabilities, and analyzes the result. It provides a product that lets an organization prepare for attacks, prioritize investments, and manage vulnerability risks based upon an understanding of their assets, their vulnerabilities, what threats correlate to the vulnerabilities, and what risks they have as a result. The cyber terrain analysis gets you to the risk mitigation stage.

As previously noted, cyber terrain is the combination of the physical devices and logical elements tied to specific functions and locations. Devices like servers and routers may be key cyber terrain if it represents a single point of failure for an environment. While a device's physical location may make it appear to be key terrain, it is the device's logical location that truly decides if it is.

While the virtual cyber terrain differs from physical terrain, the idea of observing cyber terrain is crucial. Much like when military scouts track physical terrain, cyber actors use commonly available tools to scan a network. Internal teams should routinely conduct scans to know who is on the network and to establish both what they are doing and whether it is appropriate. Friendly and hostile external actors conduct similar scans.

External and internal scans will generally yield entirely different results. Attackers can hide their intent in the routine data traffic, the network's "noise," to confuse defenders. Defenders can establish traps, creating zones within their networks to pull in attackers to establish what they are doing, how they are doing it, locations to draw in attackers to establish what and how they are trying to do, and to cause them to waste their time and effort on something of no value.

Cover, as protection from hostile activities, is often provided by firewalls and other access control systems that stop attackers from reaching specific hosts while concurrently defending those systems. Intrusion prevention systems are devices or applications that monitor for malicious behavior. These systems can be used to cover by stopping specific networks traffic. Firewalls and intrusion prevention systems provide cover, these devices do not provide concealment, the internal networks can still be observed by the attackers. Further, it cannot be assumed that the monitoring systems themselves have not been compromised.

In traditional military terms, *concealment* protects an individual or organization from observation. Cyber concealment applies to both the attacker and the defender.

Cyber attackers use concealment to avoid detection. Attackers use obfuscation techniques to reduce the potential for the firewalls and intrusion prevention systems to recognize them. They use concealment to hide the presence of either malicious code or advanced persistent threat code—the code that resides for an extended period within an environment monitoring activities, gathering data, or waiting to execute a destructive command.

For defenders, concealment through obfuscation takes a variety of forms, with technologies that employ deception to misdirect intruders and disrupt their activities. These efforts delay attackers, forcing them to spend more time and effort figuring out what is real and whether to proceed with an attack. Data obfuscation masks data, scrambling it to prevent unauthorized access. Network obfuscation involves changing data paths and routes in real time.

In cyber conflict, *obstacles* limit the ability of an attacker to freely move about in the cyber environment. Through policy and technologies, organization can place physical or logical barriers to prevent attackers from reaching their objective. Cyber "obstacles can include physical separation, access control lists, firewalls, or other devices that restrict the flow of data."

Further, the policies and technologies used to create a cyber obstacle can also provide cover. While the devices can be seen, policies and technologies can prevent them from being engaged.

SUN TZU said

1. In respect to the employment of troops, ground may be classified as dispersive, frontier, key, communicating, focal, serious, difficult, encircled, and death.

2. When a feudal lord fights in his own territory, he is in dispersive ground.

3. When he makes but a shallow penetration into enemy territory he is in frontier ground.

4. Ground equally advantageous for the enemy or me to occupy is key ground.

5. Ground equally accessible to both the enemy and me is communicating.

6. When a state is enclosed by three other states its territory is focal. He who first gets control of it will gain the support of All-under-Heaven, the entire Empire.

7. When the army has penetrated deep into hostile territory, leaving far behind many enemy cities and towns, it is in serious ground.

8. When the army traverses mountains, forest, precipitous country, or marches through defiles, marshland, or swamps, or any place where the going is hard, it is in difficult ground.

9. Ground to which access is constricted, where the way out is tortuous, and where a small enemy force can strike my larger one is called "encircled."

10. Ground in which the army survives only if it fights with the courage of desperation is called "death."

11. And therefore, do not fight in dispersive ground; do not stop in the frontier borderlands.

12. Do not attack an enemy who occupies key ground; in communicating ground do not allow your formations to become separated.

13. In focal ground, ally with neighboring states; in deep ground, plunder.

14. In difficult ground, press on; in encircled ground, devise stratagems; in death ground, fight.

15. In dispersive ground I would unify the determination of the army.

16. In frontier ground I would keep my forces closely linked.

17. In key ground I would hasten up my rear elements.

18. In communicating ground I would pay strict attention to my defenses.

19. In focal ground I would strengthen my alliances.

20. In serious ground I would ensure a continuous flow of provisions.

21. In difficult ground I would press on over the roads.

22. In encircled ground I would block the points of access and egress.

23. In death ground I could make it evident that there is no chance of survival. For it is the nature of soldiers to resist when surrounded; to fight to the death when there is no alternative, and when desperate to follow commands implicitly.

24. The tactical variations appropriate to the nine types of ground, the advantages of close or extended deployments, and the principles of human nature are matters the general must examine with the greatest care.

25. Anciently, those described as skilled in war made it impossible for the enemy to unite his van and his rear; for his elements both large and small to mutually co-operate; for the good troops to succor the poor and for superiors and subordinates to support each other.

26. When the enemy's forces were dispersed they prevented him from assembling them; when concentrated, they threw him into confusion.

27. They concentrated and moved when it was advantageous to do so; when not advantageous, they halted.

28. Should one ask: "How do I cope with a well-ordered enemy host about to attack me?" I reply: "Seize something he cherishes and he will conform to your desires."

29. Speed is the essence of war. Take advantage of the enemy's unpreparedness; travel by unexpected routes and strike him where he has taken no precautions.

30. The general principles applicable to an invading force are that when you have penetrated deeply into hostile territory your army is united, and the defender cannot overcome you.

31. Plunder fertile ground to supply the army with plentiful provisions.

32. Pay heed to nourishing the troops; do not unnecessarily fatigue them. Unite them in spirit; conserve their strength. Make unfathomable plans for the movements of the army.

33. Throw the troops into a position from which there is no escape and even when faced with death they will not flee. For if prepared to die, what can they not achieve? Then officers and men together put forth their utmost efforts. In a desperate situation they fear nothing; when there is no way out they stand firm. Deep in a hostile land they are bound together and there, where there is not alternative, they will engage the enemy in hand to hand combat.

34. Thus, such troops need no encouragement to be vigilant. Without extorting their support the general obtains it; without inviting their affection he gains it; without demanding their trust he wins it.

35. My officers have no surplus wealth but not because they disdain worldly goods; they have no

expectation of long life but not because they dislike longevity.

36. On the day the army is ordered to march the tears of those seated soak their lapels; the tears of those reclining course down their cheeks.

37. But throw them into a situation where there is no escape and they will display the immortal courage of Chuan Chu and Ts'ao Kuei.

38. Now the troops of those adept in war are used like the "Simultaneously Responding" snake of Mount Ch'ang. When struck on the head its tail attacks; when struck on the tail, its head attacks, when struck in the center both head and tail attack.

39. Should one ask: "Can troops be made capable of such instantaneous co-coordination?" I replied, "They can." For, although the men of Wu and Yueh mutually hate one another, if together in a boat tossed by the wind they would co-operate as the right hand does with the left.

40. It is thus not sufficient to place one's reliance on hobbled horses or buried chariot wheels.

41. To cultivate a uniform level of valor is the object of military administration. And it is by proper use of the ground that both shock and flexible forces are used to the best advantage.

42. It is the business of a general to be serene and inscrutable, impartial and self-controlled.

43. He should be capable of keeping his officers and men in ignorance of his plans.

44. He prohibits superstitious practices and so rids the army of doubts. Then until the moment of death there can be no troubles.

45. He changes his methods and alters his plans so that people have no knowledge of what he is doing.

46. He alters his camp-sites and marches by devious routes and thus makes it impossible for others to anticipate his purpose.

47. To assemble the army and throw it into desperate position is the business of the general.

48. He leads the army deep into hostile territory and there releases the trigger.

49. He burns his boats and smashes his cooking pots; he urges the army on as if driving a flock of sheep, now in one direction, now in another, and none knows where is going.

50. He fixes a date for rendezvous and after the troops have met, cuts off their return route just as if he were removing a ladder from beneath them.

51. One ignorant of the plans of the neighboring states cannot prepare alliances in good time; if ignorant of the conditions of mountains, forests, dangerous defiles, swaps and marshes, he cannot conduct the march of an army; if he fails to make use of native guides he cannot gain the advantages

of the ground. A general ignorant of even one of these three matters is unfit to command the armies of a Hegemonic King.

52. Now when a Hegemonic King attacks a powerful state he makes it impossible for the enemy to concentrate. He overawes the enemy and prevents his allies from joining him.

53. It follows that he does not contend against powerful combination nor does he foster the power of other states. He relies for the attainment of his aims on his ability to overawe his opponents. And so he can take the enemy's cities and overthrow the enemy's state.

54. Bestow reward without respect to customary practice; publish orders without respect to precedent. Thus you may employ the entire army as you would one man.

55. Set the troops to their tasks without imparting your designs; use them to gain advantage without revealing the dangers involved. Throw them into a perilous situation and they survive; put them in death ground and they will live. For when the army is place in such a situation it can snatch victory from defeat.

56. Now the crux of military operations lies in the pretense of accommodating one's self to the designs of the enemy.

57. Concentrate your forces against the enemy and from a distance of a thousand *li* you can kill his

general. This is described as the ability to attain one's aim in an artful and ingenious manner.

58. On the day the policy to attack is put into effect, close the passes, rescind the passports, have no further intercourse with the enemy's envoys and exhort the temple council to execute the plans.

59. When the enemy presents an opportunity, speedily take advantage of it. Anticipate him in seizing something he values and move in accordance with a date secretly fixed.

60. The doctrine of war is to follow the enemy situation in order to decide on battle.

61. Therefore at first be shy as a maiden. When the enemy gives you an opening be swift as a hare and he will be unable to withstand you.

Cyber Corollaries

1.   In respect to the employment of troops, ground may be classified as dispersive, frontier, key, communicating, focal, serious, difficult, encircled, and death.

Following the principles of "knowing yourself," and "knowing the enemy," Sun Tzu spends a great deal of time discussing how to best know the environment, the terrain, in which an organization operates. As previously noted, *key cyber terrain* is "those physical and logical elements of the domain that enable mission essential warfighting functions." While like the purely physical terrain described by Sun Tzu, cyber terrain has some significant differences due to its simultaneous physical and virtual nature. It is the combination of those natures that makes identification of and operations around cyber terrain more difficult than the other domains—land, sea, air, and space.

Cyber terrain is tied to both physical and logical locations. A hardware device, such as a router, switchsz, cable, or intrusion detection system, may be key cyber terrain if it represents a single point of failure for an environment. The device's logical location is the crucial aspect of target identification and terrain analysis. With software-defined network and multiple connection paths, the terrain features can dramatically shift in milliseconds with no apparent change in the physical configuration.

24.   The tactical variations appropriate to the nine types of ground, the advantages of close or extended deployments, and the principles of human nature are matters the general must examine with the greatest care.

Discovery and classification allows an organization to see everything that connects to the environment and understand how it connects. Correlation allows the organization to "see" behaviors, to understand that a threat in one place links to suspicious behavior elsewhere.

> 28.   Should one ask: "How do I cope with a well-
> ordered enemy host about to attack me?" I reply: "Seize
> something he cherishes and he will conform to your
> desires."

Every cyber actor values something and has a weakness. Considering Sun Tzu's admonishment to seize something that an attacker values, organizations need to engage in active discouragement. If the attacker values time, an organization can install intrusion protection or obfuscation systems to cause the attacker to expend additional time. If the attacker is financially motivated, an organization and its alliances can seek to prevent the banking community from conducting transactions for the attacker. If the attacker is politically motivated, an organization can shed light and transparency on the situation and direct psychological responses at the mind and moral of the attacker.

> 29.   Speed is the essence of war. Take advantage of
> the enemy's unpreparedness; travel by unexpected routes
> and strike him where he has taken no precautions.

This stanza includes two complementary ideas, both serving as cautionary notes in the protection of the cyber environment. First, cyber attacks happen quickly. Cyber's nature makes it possible for attackers to almost instantaneously create virtual armies of bots and machines with a global operational reach. Cyber attacks can result in the disruption of operations, financial losses, identity theft, or damaged reputation. To rapidly respond, it is important for organization to have a response plan ready that describes how they will react to the cyber attack. This response plan should include a detailed assessment of the risks and methodology by which the response team will secure the information systems network to contain the attack.

Second, as previously noted, cyber attackers look for weakness and attack an organization's vulnerable points. As part of the recognize-remediate-respond approach, organizations need to fully understand their operational and cyber operational environments. By identifying their own weaknesses, they can better prepare for the time when an attacker seeks to exploit those flaws.

XII: Attacks

The Stage

Like the Warring States period's constant strife, the current state of cyber conflict is one in which organizations are either under attack, about to be attacked, or have just been attacked. Assaults are constant and continual.

During the Warring States period, fire attacks were the deadliest form of warfare. Fire and various types of incendiary devices were frequently used against enemy structures and territory. Sun Tzu used the most destructive attack he knew of, fire attacks, as a framework for discussing techniques to destroy an opponent.

This chapter examines the five fire attack targets and five types of environmental attacks. It also addresses the appropriate responses to these attacks. It concludes with a specific caution not to attack based upon emotions and a generic warning about the use of force in general.

A cyber attack, also referred to as a computer network attack, is the deliberate, malicious exploitation of or use of computer systems, computer networks, computer infrastructure, and other related devices to gain, use, damage, alter, or steal from the system achieved by hacking and placing malicious code into the susceptible systems; anonymous parties usually propagate these actions. Using the analysis process, organizations can plan for the different types of attacks that may be used again them and prepare responses.

As noted in chapter 5, "Agency," cyber actions can be described in a variety of manners: effect, extent, and impact. The first category is the methods with some specific effects in mind (nuisance, disruptive, disinformation, crime, espionage, and destructive). The *extent* of a cyber attack can be characterized in terms of scale and time frame (isolated, bounded in time and scope, persistent, and large-scale).

Intended impact can be identified in several ways—in terms of the effects on services provided by cyber resources, on the adversary's activities, or on risk and/or resilience (terminate, restrict, alter, and observe).

Cyber attacks may result in disruption of operations, financial losses, identity theft, damaged reputation, and the closure of business. Consequently, it is important for organizations to have in place a response plan that delineates how they may respond to a cyber attack.

The first step in achieving preparedness against cyber attacks is to have a detailed risk-management plans that include the baseline, an assessment of the risk of cyber attacks, and a response plan. When a cyber attack occurs, the organization should firstly mobilize an incident response team, which will include members drawn from the information technology department and other departments of the organization to manage the crisis effectively. The incident response team is made up of technical professionals to contain and investigate the cyber attack and professionals from other departments, such as the personnel department (to handle personnel issues that may arise), legal department, public relations, and members of the executive management department.

The response team will understand the impact and secure the network and systems. They need to contain the attack to ensure that it is not ongoing, while concurrently undertaking activities that foster business continuity. Once the attack has been contained and the operations of the firm are going on, the technical team should then investigate the source of attack, the agenda of the attackers, the extent of damage and impact on business operations.

In reaction to a cyber attack, after the organization has controlled the attack and begun investigations, the public relations team should communicate with all the stakeholders promptly in accordance with the laid-out crisis communication and public relations plan. Finally, the organization should take redress actions, such as incurring liability and making payments to those that might have been affected by the attack, and ensure compliance with all legal and regulatory requirements.

Cyber attacks will happen, and then will happen again. By understanding themselves, their disposition, and their weaknesses, and understanding their adversaries' motivations and tactics, organizations can prepare themselves and their environments to recognize and remediate attacks and to continue their operations.

SUN TZU said

1. There are five methods of attacking by fire. The first is to burn personnel; the second, to burn stores; the third, to burn equipment; the fourth, to burn arsenals; and the fifth, to use incendiary missiles.

2. To use fire, some medium must be relied upon.

3. Equipment for setting fires must always be at hand.

4. There are suitable times and appropriate days on which to raise fires.

5. "Times" means when the weather is scorching hot; "days" means when the moon is Sagittarius, Alpharatz, I, of Chen constellations, for these are days of rising winds.

6. Now in fire attacks one must respond to the changing situation.

7. When fire breaks out in the enemy's camp immediately co-ordinate your action from without. But if his troops remain calm bide your time and do not attack.

8. When the fire reaches its heights, follow up if you can. If you cannot do so, wait.

9. If you can raise fires outside the enemy camp, it is not necessary to wait until they are started inside. Set fires at suitable times.

10. When fires are raised up-wind do not attack down-wind.

11. When the wind blows during the day it will die down at night.

12. Now the army must know the five different fire-attack situations and be constantly vigilant.

13. Those who use fire to assist their attack are intelligent; those who use inundations are powerful.

14. Water can isolate an enemy but cannot destroy his supplies or equipment.

15. Now to win battles and take your objectives, but to fail to exploit these achievements is ominous and may be described as "wasteful delay."

16. And therefore it is said that enlightened rulers deliberate upon the plans and good generals execute them.

17. If not in the interests of the state, do not act. If you cannot succeed, do not use troops. If you are not in danger, do not fight.

18. A sovereign cannot raise an army because he is enraged, nor can a general fight because he is resentful. For while an angered man may again be happy, a resentful man again be pleased, a state that has perished cannot be restored, nor can the dead be brought back to life.

19. Therefore, the enlightened rule is prudent and the good general is warned against rash action. Thus the state is kept secure and the army preserved.

Cyber Corollaries

1.   There are five methods of attacking by fire. The first is to burn personnel; the second, to burn stores; the third, to burn equipment; the fourth, to burn arsenals; and the fifth, to use incendiary missiles.

Sun Tzu used fire, the most destructive attack known, to describe offensive actions. In military jargon, *fires* are "the use of weapon systems to create specific lethal or nonlethal effects on a target."

Cyber fires are the use of technology to create specific lethal or nonlethal effects on a physical or virtual target. From the effects (nuisance, disruption, disinformation, criminal, espionage, destruction) to the extent (isolated, bounded, persistent) and the intended impact (observe, restrict, alter, terminate), organizations need to be aware of the type of cyber fires that may be directed at them and be prepared to take the appropriate response. Some cyber fires can be mitigated through industrialized remediate and user training. Other fires can be maneuvered through using robust architectures.

XIII: Information Gathering

The Stage

The title's Chinese character means "the space between" two objects, or divide. The traditional translations are "Spies" or "Using Spies." These translations lead the cursory reader down the wrong path. For Sun Tzu, the topic is that of gathering information. Accurate, actionable information allows leaders to make fully informed decisions. Inaccurate biases and filtered information leads organizations astray. The chapter discusses methods of gathering and managing information and the necessity for doing so.

The collection of information is as old as time. Sun Tzu advocates dispatching some men to scout the ground to find the best routes, to forage, or to gather water. Other men are sent to find the enemy positions. Other are sent to provoke the adversary into declaring themselves. Still others are sent to intercept the messengers that carry the communications of the king. Spies are sent into the enemy camp and cities to seek out the strength of his positions, to penetrate the council of the leader, to report on his intentions, and ideally to influence them.

The need for information is a common feature among individuals, organizations, and states. When faced with almost overwhelming cyber risks and uncertainty, organizations face a "knowledge problem," an inability to discern the truth.

Cybersecurity information gathering allows them to resolve that problem. Information gathering enables the stakeholders to understand and remediate their weak points.

Cybersecurity information gathering involves both the use of internal and third-party sources. These sources allow organizations to better understand who is targeting them, the methods they are using, their techniques, and their targets. Understanding the threat actors, their methods, and how to detect or prevent attacks assists organizations in the process-ing of shaping both policies and actions to mitigate the effects (Gurjar, 2013). If cyber intelligence is used correctly, it can assist in the detection of attacks before they take place and provide indicators of actions that should be taken dur-ing attacks.

SUN TZU said

1. Now when an army of one hundred thousand is raised and dispatched on a distant campaign the expenses borne by the people together with the disbursements of the treasury will amount to a thousand pieces of gold daily. There will be continuous commotion both at home and abroad, people will be exhausted by the requirements of transport, and the affairs of seven hundred thousand households will be disrupted.

2. One who confronts his enemy for many years in order to struggle for victory in a decisive battle yet who, because he begrudges rank, honors and a few hundred pieces of gold, remains ignorant of his enemy's situation, is completely devoid of humanity. Such a man is no general; no support to his sovereign; no master of victory.

3. Now the reason the enlightened prince and the wise general conquer the enemy whenever they move and their achievements surpass those of ordinary men is foreknowledge.

4. What is called "foreknowledge" cannot be elicited from spirits, nor from gods, nor by analogy with past events, nor from calculations. It must be obtained from men who know the enemy situation.

5. Now there are five sorts of secret agents to be employed. These are native, inside, doubled, expendable, and living.

6. When these five types of agents are all working simultaneously and none knows their method of

operation, they are called "The Divine Skein" and are the treasures of the sovereign.

7. Native agents are those of the enemy's people whom we employ.

8. Inside agents are enemy officials whom we employ.

9. Double agents are enemy spies whom we employ.

10. Expendable agents are those of our own spies who are deliberately given fabricated information.

11. Living agents are those who return with information.

12. Of all those in the army close to the commander none is more intimate than the secret agent; of all rewards none more liberal than those given to secret agents; of all matters, none is more confidential than those relating to secret operations.

13. He who is not sage and wise, humane and just, cannot use secret agents. And he who is not delicate and subtle cannot get the truth out of them.

14. Delicate indeed! Truly delicate! There is no place where espionage is not used.

15. If plans relating to secret operations are prematurely divulged the agent and all those to whom he spoke of them shall be put to death.

16. Generally in the case of armies you wish to strike, cities you wish to attack, and people you wish to

assassinate, you must know the names of the garrison commander, the staff officers, the ushers, the gate keepers, and the bodyguards. You must instruct your agents to inquire into these matters in minute detail.

17. It is essential to seek out enemy agents who have come to conduct espionage against you and to bribe them to serve you. Give them instructions and care for them. Thus doubled agents are recruited and used.

18. It is by means of the doubled agent that native and inside agents can be recruited and employed.

19. And it is by this means that the expandable agent, armed with false information, can be sent to convey it to the enemy.

20. It is by this means also that living agents can be used at appropriate times.

21. The sovereign must have full knowledge of the activities of the five sorts of agents. This knowledge must come from the doubled agents, and therefore it is mandatory that they be treated with the utmost liberality.

22. Of old, the rise of Yin was due to I Chih, who formerly served the Hsia; the Chou came to power through Lu Yu, a servant of the Yin.

23. And therefore only the enlightened sovereign and the worthy general who are able to use the most intelligent people as agents are certain to achieve great things. Secret operations are essential in war; upon them the army relies to make its every move.

Cyber Corollaries

4.   What is called "foreknowledge" cannot be elicited from spirits, nor from gods, nor by analogy with past events, nor from calculations. It must be obtained from men who know the enemy situation.

By "foreknowledge," Sun Tzu means the probability of future events, a probability that can only be determined by both the possession of information and its proper interpretation. Gathered information has no value unless it accurately analyzed and contextually applied.

From the above, it is apparent that the need for information is a common feature among organizations, individuals, states, groups, and societies. As noted, when faced with cyber risks, uncertainty, and feelings of insecurity, people face a "knowledge gap." Cybersecurity informational gathering is important since it relieves, at least in part, the knowledge gap and enables the stakeholders to recognize-remediate-respond.

Organizations should understand the importance of gathering cybersecurity information and conducting regular information security assessments. To achieve objectivity, these informational assessments should be carried out by an external service provider assisted by the internal informational technology department.

Cybersecurity informational gathering is important since it helps organizations to stay alert and to understand the internal and external security risks. If left unmanaged, these security risks can lead to successful cyber attacks.

5.   Now there are five sorts of secret agents to be employed. These are native, inside, doubled, expendable, and living.

To "know yourself" and "know your enemy," Sun Tzu's "secret agents" can best be thought of as information sources. An advantage today is the availability of a vast pool of information and numerous information sources, especially in cyber threat intelligence.

Information access, business intelligence, and their use is a key to driving competitiveness and organizational success. Business intelligence is defined as the process of collecting, analyzing, manipulating, storing, and using raw data and information to facilitate organizational operations, and these processes involve accessing and using information to optimize organizational performance and competitiveness. Cyber intelligence gathering applies this same definition to the cyber subset.

As with business intelligence, organizations should also understand the importance of cybersecurity informational gathering and conducting information security assessment regularly. Informational assessment should be carried out by an external service provider assisted by the internal informational technology department. Cybersecurity informational gathering is important since it helps organizations to stay aware of and to contextually understand their security risks.

An understanding of the threat actors, their methods, the environment, and the threat can greatly assist in the process of shaping necessary policies and actions to mitigate the effects (Gurjar, 2013). If cyber intelligence is used correctly, it can detect attacks before they take place and provide indicators of actions that should be taken during attacks.

Information gathering is the most important task for the attackers. This is because it enables the attackers to understand their opponent's interests, operations, tactics, disposition, terrain, and targets.

As with Sun Tzu's secret agents, there are multiple ways to gather cyber threat intelligence. Source selections involves two questions:

1. Is the information actionable within the context of the organization's environment?

2. Does the information assist the organization in achieving its long-term objectives?

There are five main categories of threat intelligence sources that an organization can choose from:

1. Internal: Internal sources are both technical and human. Technical sources include system and sensor log files and vulnerability assessments. Human sources are all the organization's team members who could provide insights into what is of value, their training, and what they perceive are the threats.

2. Alliance: In some cases, organizations have entered industry partnerships and consortiums; these are their alliances. Alliances should be sharing cyber threat information among its participants.

3. Vendors: The infrastructure is bought from and built upon products, services, and technologies supplied by outside vendors. These vendors should be providing organizations cyber information regarding the goods they have provided.

4. Government: As a citizen service, many governments provide best practices and threat information that organizations can avail themselves of.

5. Private sources: An industry has developed around cyber threat intelligence gathering and distribution. Organizations can purchase this third-party data to gain external insights regarding their risk profile.

In the end, organizations need to find their own "secret agents" and establish which sources of cyber information best fit the organization.

CONCLUSION

> For to win one hundred victories in one hundred battles is not the acme of skill. To subdue the enemy without fighting is the acme of skill.

> —*The Art of War, I:3*

"To subdue the enemy without fighting is the acme of skills" is one of the most quoted lines from *The Art of War*. In cyber conflict, there are hundreds of thousands of battles daily. The conflict is continuous, unrestricted, and omnidirectional. There is no apparent end in sight.

The victors in this conflict will not be those with the best technology or with the best people. Success will be for those who best leverage the cyber environment to confuse, deceive, and control the adversary and continue their operations in the face of cyber attack.

It comes down to assuming organizations are always under attack or at least always being monitored. Once organizations become aware of their environments and its vulnerabilities, they can act to prevent their environments and vulnerabilities from being using against them. Once an organization accepts the fact it is likely to be hacked, it can take advantage of knowing enough about its own environment to actively discourage attackers.

The prevention of attacks is not easy; there is no such thing as a silver bullet. Even with the recognize-remediate-respond approach, few organizations have the resources to stay on top of all the digital threats.

Further, organizations should recognize that there is never a way to completely eliminate risk. The only risk-free information technologies are completely isolated technologies, a useless approach in today's world.

It is the context of change, conflict, and understanding the environment that allows *The Art of War* and Sun Tzu's principles to be used as a planning framework to ensure success in cyber conflict.

SELECT BIBLIOGRAPHY

I have attempted to include all the pieces that influenced this book.

My apologies for any omissions; the omissions were unintentional.

Adler, E. "Seizing the Middle Ground: Constructivism in World Politics." *European journal of International Relations* 3, no.3 (1997).

Applegate, Scott D. "The Principle of Maneuver in Cyber Operations." Fourth International Conference on Cyber Conflict, 2012.

Ascher, William, and Natalia Mirovitskaya. *Development Strategies and Inter-Group Violence*. New York: Palgrave Macmillan, 2015.

Austin, G. *Cyber Policy in China*. Cambridge, UK: Polity Press, 2014.

Bagchi, K., and G. Udo. "An Analysis of the Growth of Computer and Internet Security Breaches." *Communications of the Association for Information Systems* 12, no. 1 (2003).

Basu, E. "What Is a Penetration Test and Why Would I Need One for My Company?" October 13, 2013. Retrieved from http://www.forbes.com/sites/ericbasu/2013/10/13/what-is-a-penetration-test-and-why-would-i-need-one-for-my-company/#2e143ba342da.

Bodeau, Deborah, Richard Graubart, and William Heinbockel. *Mapping the Cyber Terrain*. MITRE TECHNICAL REPORT MTR1 3 04 3 3. McLean, VA: MITRE Corporation, 2013.

Brenner, Joel. *America the Vulnerable*. New York: Penguin, 2011.

Bronfen, E. *Specters of War: Hollywood's Engagement with Military Conflict*. New Brunswick, NJ: Rutgers University Press, 2012.

Brück, Tilman, Patricia Justino, Philip Verwimp, Alexandra Avdeenko, Alexandra, and Andrew Tedesco. "Measuring Violent Conflict in Micro-Level Surveys: Current Practices and Methodological Challenges." *World Bank Research Observer* 31, no. 1 (2016).

Burgess-Proctor, A., J. Patchin, and S. Hinduja. "Cyberbullying and Online Harassment: Conceptualizing the Victimization of Adolescent Girls." In *Female Crime Victims: Reality Reconsidered*, edited by Venessa Garcia and Janice E. Clifford. Upper Saddle River, NJ: Prentice Hall, 2009.

Burgman, P. "Securing Cyberspace: China Leading the Way in Cyber Sovereignty." May 18, 2016. Retrieved from http://thediplomat.com/2016/05/securing-cyberspace-china-leading-the-way-in-cyber-sovereignty/.

Capgemini. "Anomalous Behaviour Detection." Retrieved December 5, 2016. https://www.capgemini.com/insightsdata/insights/anomalousbehaviordetection.

Chambers, John R., Barry R. Schlenker, and Brian Collisson. "Ideology and Prejudice: The Role of Value Conflicts." *Psychological Science* 24, no. 2 (2013).

Cloke, K., and J. Goldsmith. *Resolving Conflicts at Work: Ten Strategies for Everyone on the Job*. San Francisco: Jossey-Bass, 2011.

Denning, P. J., and D. E. Denning. "Discussing Cyber Attack." *Communications of the ACM* 53, no. 9 (2010).

Elazari, K. "How to Survive Cyberwar." *Scientific American* 312, no. 4 (2015).

Finn, T. T. *America at War: Concise Histories of US Military Conflicts from Lexington to Afghanistan*. New York: Berkley Caliber, 2014.

Florian, C. "5 Benefits of Automating Patch Management." *GFI Blog*. Retrieved January 30, 2017, from https://techtalk.gfi.com/5-benefits-automating-patch-management/.

Gelpi, C., P. Feaver, and J. A. Reifler. *Paying the Human Costs of War: American Public Opinion and Casualties in Military Conflicts*. Princeton, NJ: Princeton University Press, 2009.

Gladwell, M. *David and Goliath*. Boston: Little, Brown, 2013.

Gourley, S. K. "Cyber Sovereignty." In *Conflict and Cooperation in Cyberspace: The Challenge to National Security*, edited by Panayotis A. Yannakogeorgos and Adam B Lowther. Boca Raton, FL: Taylor & Francis, 2013.

Gurjar, C. "Network Intelligence Gathering." 2013. Retrieved from http://resources.infosecinstitute.com/network-intelligence-gathering/.

Gutteridge, Joyce A. C. "The Geneva Conventions of 1949." *British Yearbook of International Law* 26 (1949).

Hacker, B. C. "Military Institutions, Weapons, and Social Change: Toward a New History of Military Technology." *Technology and Culture* 35, no. 4 (1994).

Herzog, S. "Revisiting the Estonian Cyber Attacks: Digital Threats and Multinational." *Journal of Strategic Security* 4, no. 2 (2011).

Heumann, Silke, and Jan Willem Duyvendak. "When and Why Religious Groups Become Political Players." In

Players and Arenas, edited by James M. Jasper and Jan Willem Duyvendak. Amsterdam, Netherlands: Amsterdam University Press, 2015.

Hollywood, John, Diane Snyder, Kenneth McKay, John Boon. *Out of the Ordinary: Finding Hidden Threats by Analyzing Unusual Behavior*. Santa Monica, CA: RAND Corporation, 2004.

Householder, A., K. Houle, and C. Dougherty. "Computer Attack Trends Challenge Internet Security." *Computer* 35, no. 4 (2002).

Jennex, M. E. "Modeling Emergency Response Systems." In *System Sciences, 2007. HICSS 2007*. IEEE, 2007.

Jensen, E. T. "Sovereignty and Neutrality in Cyber Conflict." *Fordham International Law Journal* 35, no. 3 (2011).

Kanwal, Gurmeet. "China's Emerging Cyber War Doctrine." *Journal of Defence Studies* 3, no. 3 (July 2009).

Katherine, D. S. "Converging, Pervasive Technologies: Chronic and Emerging Issues and Policy Adequacy Assistive Technology." *Official Journal of RESN* 20, no. 3 (2008).

Kowalski, R. M., and S. P. Limber. "Psychological, Physical, and Academic Correlates of Cyberbullying and Traditional Bullying." *Journal of Adolescent Health* 53, no. 1 (2013).

Kugler, R. L. "Deterrence of Cyber Attacks." In *Cyberpower and National Security*, edited by Franklin Kramer, Stuart H. Starr, and Larry Wentz. Dulles, VA: Potomac Books, 2009.

Kugler, R. L. *US Military Strategy and Force Posture for the Twenty-First Century*. Santa Monica, CA: RAND Corporation, 1994.

Kupcikas K. "The Importance of Intelligence to International Security." 2013. Retrieved from http://www.e-ir.info/2013/11/08/importance-of-intelligence-to-international-security/.

Lai, Robert. "Analytic of China Cyber Warfare." RSA Conference, 2013.

Larsen, P., A. Homescu, S. Brunthaler, and M. Franz. "SoK: Automated Software Diversity." In *Security and Privacy (SP), 2014.* IEEE, 2014.

Larson, Quincy. "How to Encrypt Your Entire Life in Less Than an Hour." November 9, 2016. https://medium.freecodecamp.com/tor-signal-and-beyond-a-law-abiding-citizens-guide-to-privacy-1a593f2104c3#.1zvpmkmmr.

Liang, Qiao, and Wang Xiangsui. *Unrestricted Warfare: China's Master Plan to Destroy America.* Panama City, Panama: Pan American Publishing, 2002.

Lie, E., R. Macmillan, and R. Keck. "Cybersecurity: The Role and Responsibilities of an Effective Regulator." Paper presented at 9th ITU Global Symposium for Regulators, Beirut, Lebanon, November 2009.

Lewis, J. A. "Sovereignty and the Role of Government in Cyberspace." *Brown Journal of World Affairs* 16, no.2 (2010).

Limbago, A. "The Global Push for Cyber Sovereignty Is the Beginning of Cyber Fascism." *The Hill.* 2016. http://thehill.com/blogs/congress-blog/technology/310382-the-global-push-for-cyber-sovereignty-is-the-beginning-of.

Lindstrom, P. "A Patch in Time: Considering Automated Patch Management." February 2004. Retrieved February 1, 2017, from http://searchsecurity.techtarget.com/A-Patch-in-Time-Considering-automated-patch-management-solutions.

Loshin, D. *Business Intelligence: The Savvy Manager's Guide.* Waltham, MA: Morgan Kaufmann, 2012.

Lowery, J. "Penetration Testing: The Third Party Hacker." SANS GIAC practical exam. http://www.sans.org/rr/paper.php.

McAfee. "The Economic Impact of Cybercrime and Cyber Espionage." July 2013.

Mell, P., T. Bergeron, and D. Henning. "NIST Special Publication 800-40: Creating a patch and vulnerability management program." *Recommendations of the National Institute of Standards and Technology.* Gaithersburg, MD: National Institute of Standards and Technology, 2005.

Moore, Alexis, and Laurie J. Edwards. *Cyber Self-Defense.* Guildford, CT: Lyons Press, 2014.

Nance, Richard. "Sun Tzu and the Art of Defensive Tactics." Retrieved October 30, 2016, from http://www.officer.com/article/10249379/sun-tzu-and-the-art-of-defensive-tactics.

Newell, C. *The Framework of Operational Warfare.* Abingdon, UK: Routledge, 2003.

Paganini, Pierluigi. "Cybersecurity: Red Team, Blue Team and Purple Team." Retrieved December 13, 2016, from http://securityaffairs.co/wordpress/49624/hacking/cyberredteamblueteam.

Peltsverger, S., and G. Zheng. "Enhancing Privacy Education with a Technical Emphasis in IT Curriculum." *Journal of Information Technology Education: Innovations in Practice* 15 (2016).

Pettersson, Therése, and Peter Wallensteen. "Armed Conflicts, 1946–2014." *Journal of Peace Research* 52, no.4 (2015).

Putnam, Robert. *Bowling Alone* New York: Simon & Schuster, 2001.

Qi, Y., X. Mao, Y. Lei, Z. Dai, and C. Wang. "The Strength of Random Search on Automated Program Repair." In *Proceedings of the 36th International Conference on Software Engineering.* New York: ACM, 2014.

Ralph, D. "Cambridge Centre for Risk Studies Trevor Maynard Head, Exposure Management & Reinsurance Performance Management, Lloyd's." 2015. Retrieved from https://www.lloyds.com/cityriskindex/files/8771-City-Risk-Executive-summary-AW.pdf.

Raymond, David, Gregory Conti, Tom Cross, and Michael Nowatkoski. "Key Terrain in Cyberspace: Seeking the High Ground."

RSIS. *Cybersecurity: Emerging Issues, Trends, Technologies, and Threats in 2015 and Beyond.* Singapore: S. Rajaratnam School of International Studies, 2016.

Ryan, J., ed. *Leading Issues in Cyber Warfare and Security.* Sonning Common, UK: ACPIL, 2015.

Sabherwal, R., and I. Becerra-Fernandez. *Business Intelligence.* Hoboken, NJ: Wiley, 2009.

Samuels, M. *Doctrine and Dogma: German and British Infantry Tactics in the First World War.* New York: Greenwood Publishing, 1992.

Sechrist, Michael. *Cyberspace in Deep Water: Protecting Undersea Communication Cables.* Cambridge, MA: Harvard Kennedy School, 2010.

Sekhar, G. *Business Policy and Strategic Management*. New York: I. K. International, 2009.

Sienkiewicz, Henry J. "Inside the mind of a Designated Approving Authority." Defense Systems. October 10, 2013. https://defensesystems.com/articles/2013/10/10/designated-approving-authority.aspx.

Smith, E. K., E. T. Barr, C. Le Goues, and Y. Brun. "Is the Cure Worse Than the Disease? Overfitting in Automated Program Repair." In *Proceedings of the 2015 10th Joint Meeting on Foundations of Software Engineering*. New York: ACM, 2015.

Smith, Rupert. *The Utility of Force: The Art of War in the Modern World*. New York: Vintage, 2008.

Snyder, Glenn Herald, and Paul Diesing. *Conflict Among Nations: Bargaining, Decision Making, and System Structure in International Crises*. Princeton, NJ: Princeton University Press, 2015.

SteelCloud. ConfigOS white paper. June 1, 2015.

Sushil Jajodia, Sushil, Steven Noel, Pramod Kalapa, Massimiliano Albanese, and John Williams. "Mission-Centric Cyber Situational Awareness." CyVision white paper.

Todd, B. "The Top 3 Benefits of Automated Patch Management for IT." Paranet.com. 2017. Retrieved January 30, 2017, from http://www.paranet.com/blog/bid/132538/The-top-3-benefits-of-Automated-Patch-Management-for-IT.

Veris Group. "Adaptive Red Team Tactics—Compliance." Retrieved December 13, 2016, from https://www.verisgroup.com/adaptiveredteamtactics/.

Wei, L. "Persisting in Respect for the Principle of Cyber Sovereignty, Promoting the Construction of a Community of Common Destiny in Cyberspace." March 2, 2016. Retrieved from https://chinacopyrightandmedia.wordpress.com/2016/03/02/persisting-in-respect-for-the-principle-of-cyber-sovereignty-promoting-the-construction-of-a-community-of-common-destiny-in-cyberspace.

White House. "US International Strategy for Cyberspace." 2011. https://www.whitehouse.gov/sites/default/files/rss_viewer/international_strategy_for_cyberspace.pdf.

Winterfeld, Steven P. "Cyber IPB." Bethesda, MD: SANS Institute, 2001.

Woodside, J. "Business Intelligence Best Practices for Success." In *International Conference on Information Management and Evaluation*. Sonning Common, UK: Academic Conferences International Limited, 2011.

Xinhua. "Why Does Cyber-Sovereignty Matter?" *China Daily*. December 16, 2015. Retrieved from http://www.chinadaily.com.cn/business/tech/2015-12/16/content_22728202.html.

Yaroslav, R. *Cyber Attacks and the Exploitable Imperfections of International Law*. Leiden, Netherlands: Brill, 2015.

Ybarra, M. L., M. Diener-West, and P. J. Leaf. "Examining the Overlap in Internet Harassment and School Bullying: Implications for School Intervention." *Journal of Adolescent Health* 41, no. 6 suppl 1 (2007).

Yeoh, W., and A. Koronios. "Critical Success Factors for Business Intelligence Systems." *Journal of Computer Information Systems* 50, no. 3 (2010).

ACKNOWLEDGMENTS

*Each friend represents a world in us, a world not possibly
born until they arrive, and it is only by this meeting
a new world is born.*

—*Anaïs Nin*

As before, it remains tough to thank everyone who helped me through this phase of my journey. My deepest gratitude goes to my family and friends. For my early readers who expressed a great deal of candor from a deep well of love. They all have inspired me, challenged me, forced me to look at myself, and helped me to grow. I am eternally grateful.

Again, my deepest appreciation to my initial readers who helped me as I struggled through this: Robert Richardson, Andrew Wonpat, Alma Miller, Dorinda Smith, Venkat Sundaram, Adam Firestone, Chris Grady, Dorinda Smith, Andrew Hiller, and Jeffrey Brady.

In order to gain other perspectives on the matters addressed, I used a variety of independent researchers through the website Fiverr. Some of the insights commissioned have incorporated into this work.

My profound thanks goes to all of those who read the galleys, put up with me, and inspired me as I wrote this: Richard and Dawn Cwirka; Ed and Barbara Cwirka; Nikki Cwirka and Perry Kloska; Austin Wineholt; Dan and Lynn Anderson; Dr. Stephanie McMillian; Julia Fischer; Anje Berger; Alicia Allen; Michael Higgs; Shelly Barber; Jennifer Augustine; Pete Tseronis; Frank Konieczny; Austin Wineholt; Lary Cohen; Scott Logan; Jason Spritzer; Linus Barloon; Gordon Tweedie; Jim Hunt; Lorraine Castro, Joanne Isham; Mark Gercenstein; Mark and Susan Orndorf; Alma Miller; Robert "RJ" Eisenhauer; Martin Gross and Kendra Altman; Kevin and Maureen Clement; Warren Coats and Victorino "Ito" Briones; Thomas

Zellars, Russ Weber, Steve Pinto; John and Geri Krause; Kathleen McBride; David MacSwain; Anthony Levensalor; David Tolson; John Kraft; Chris Grady; Andrew Wonpat; Dorinda Smith; Robert "Bob" Sacheli; Kevin Wells; Greg Dean and Gian Pertusi; Scott Mendenhall; Carter, Jennifer, Dorothy and Sarah Lane; Andy and Marlene Tucker; Dirk Keustermans; Ruben Blockx and Linde Vandermolen; Liesbet Fortje and Nico Tuytten; Karin Fortje and Rudy Blockx; Denise Keustermans and Herman Fortje; Doug Dixon, Rachel Meghan, and Rose Dixon; Daniel Schneider, Emily Hellmuth, and Ridley Schneider; Rob Lalumondier; Corey Rooney; Andrew Hiller; Florence and Mel Anderson; Nelson and Jean Ehinger; Ben, Diane, Christian, and Liz Brady; Seth June; Fr. Patrick Neary, CSC; Jason DeMoranville; Roberta and Charlie Osborn; Florence and Ed Wallace; Robert, John, Elise and Kim; Julia, Maddy, Charlotte, Grace, Kayden, Maeve, and Oliva; my parents; and Jeffrey Brady.

ABOUT THE AUTHOR

Henry is currently providing strategic advice to a select number of emerging technology companies, focused specifically on cybersecurity, data centers, and mobility.

From May 2008 until the end of June 2013, Henry was a member of the United States Senior Executive Service (SES) assigned to the Defense Information Systems Agency (DISA). While a member of the SES, he served in multiple capacities. Notably, he served as DISA's chief information officer (CIO) for two years and as the designated approving authority (DAA) for three years. From 1996 until 2008, he had a varied commercial career, primarily as the chief executive officer, chief operating officer, or chief information officer equivalent. Henry is a retired Army Reserve lieutenant colonel.

Henry holds a bachelor of arts from the University of Notre Dame and a master of science from the Johns Hopkins University. He and his partner reside in Alexandria, Virginia.

Disclaimer

As with my previous books, the views expressed here are my own and do not necessarily represent the views of any organization I am associated with.

Bulk/Trade Purchases

Please contact the publisher.

Appearances

Please contact the author.

Media Kit

A media kit may be downloaded at www.theartofcyberconflict.com.

Reviews

I love reviews and encourage readers to post them on every
forum.